How Smart Managers Create World-Class Safety, Health, and Environmental Programs

Charlotte A. Garner, CSP

American Society of Safety Engineers ❖ Des Plaines, Illinois USA

How Smart Managers Create World-Class Safety, Health, and Environmental Programs
© 2004 by Charlotte A. Garner, CSP

Published by the American Society of Safety Engineers, Des Plaines, Illinois.

No part of this publication may be reproduced, stored in a retrieval system, or transmitted by any form or by any means, electronic, mechanical, photocopying, recording, scanning, or otherwise, except as permitted under Section 107 or 108 of the 1976 United States Copyright Act, without prior written permission of the Publisher.

Limit of Liability/Disclaimer of Warranty: While the publisher and author have used their best efforts in preparing this book, they make no representations or warranties with respect to the accuracy or completeness of the contents of this book, and specifically disclaim any implied warranties of merchantability or fitness for a particular purpose. The information is provided with the understanding that the authors are not hereby engaged in rendering legal or other professional services. If legal advice or other professional assistance is required, the services of a competent professional should be sought.

Library of Congress Cataloging-in-Publication Data
Garner, Charlotte A.
    How smart managers create world-class safety, health, and environmental programs / Charlotte A. Garner.
        p.   cm.
    Includes bibliographical references and index.
    ISBN 1-885581-48-3 (alk. paper)
    1. Industrial hygiene—Management. 2. Industrial safety—Management. I. Title.
HD7261.G367   2004
658.4'08–dc22
                                                                200404365

Managing Editor: Michael Burditt, ASSE
Project Editor: Charles T. Coffin, ASSE
Copyediting, Text Design and Composition: Nancy Kaminski
Cover Design: Michael Burditt

ISBN   1-885581-48-3

Printed in the United States
10 9 8 7 6 5 4 3 2 1

Dedication

To my dear friend, Patricia Horn

Without her values, inspiration, and dedicated efforts there would be no model safety and health management system to present.

Table of Contents

	Dedication	i
	Foreword	v
	Preface	vii
Chapter 1	Challenges and Opportunities to Prevent Injuries and Illnesses	1
Chapter 2	Why Do This? The Bottom Line	19
Chapter 3	Is This Model SH&E Program for You? SH&E Management System Self-Assessment	33
Chapter 4	Where Do You Want to Be? Gap Analysis	53
Chapter 5	How to Get There From Here: The Strategic Plan	71
Chapter 6	Implement the Model SH&E Management System	89
Chapter 7	Where Are You Now? Annual Self-Evaluation of the Company's SH&E Management System	111
Chapter 8	The Two Linchpins: Leadership Commitment and Employee Involvement	129
Chapter 9	The Stakeholders: Culture and Communications	145
Chapter 10	The Never-Ending Journey: Continuing Improvement	165
Chapter 11	Epilogue: Organization of the Future	185
Appendix A	Models of Safety and Health Excellence Act of 2001	197
Appendix B	Sources of Help	201
Appendix C	Requirements for Star, Merit, Resident Contractor, Construction Industry, and Federal Agency Worksites	205
Appendix D	Onsite Evaluation Report Format	227
Appendix E	Recommended Interview Questions	255
Appendix F	Environmental Compliance Information	261
Appendix G	Program Evaluation Profile (PEP) Example	263
	Bibliography	285
	Index	289

Foreword

In the five years since the American Society of Safety Engineers published *How Smart Managers Improve Their Safety and Health Systems: Benchmarking with OSHA VPP Criteria*, which I co-authored with Patricia Horn, there have been significant changes in the workplace and in federal regulations affecting occupational ssafety and health.

On July 24, 2000, the Occupational Safety and Health Administration (OSHA) published *Revisions to the Voluntary Protection Programs to Provide Safe and Healthful Working Conditions* (Federal Register No. 65:45649-45663). The purpose of the revisions was multi-fold, but primarily as stated "to make the VPP more challenging and to raise the level of safety and health achievement expected of participants." They also "bring the VPP's basic program elements into conformity with OSHA's Safety and Health Program Management Guidelines (the Guidelines)," which were issued in the Federal Register on January 26, 1989.

Before 1989, the VPP criteria were covered by the *Voluntary Protection Programs (VPP)* adopted by OSHA in the Federal Register 29025, July 2, 1982. With the issuance of July 24, 2000, OSHA made the Guidelines of 1989 the basic reference documentation for VPP Criteria, stating, "To maintain consistency in OSHA's approach to safety and health program management, the Agency decided to reorganize the VPP criteria to conform more closely to the Guidelines."

Not only has the basic reference document changed, several program requirement changes were made in the July 24, 2000, document. "The three most notable changes are an expansion of eligibility to certain classes of worksites previously not covered by the program, increased expectations concerning the management of the safety and health of contractors' employees working at VPP sites, and a new illness reporting requirement. This last means OSHA will consider a worksite's illness experience together with its injury performance when assessing the site's level of achievement."

What does this have to do with a revision of this book? A great deal. The effort especially concerns a company's voluntary actions to improve the working conditions of American workers without seeking OSHA approval, thereby also maintaining a high level of protection from workplace hazards.

Whether a company's management decides to seek OSHA's approval for acceptance into the VPP select or just wants to take whatever safety and health steps are necessary to do what is right, the VPP management system works. More and more companies are voluntarily taking the high road to

Foreword

excellence in their SH&E management systems. In 1999 the number of VPP-approved companies was hovering around 500 after 17 years of effort. Now, at the beginning of 2004, VPP-approved companies number over 1,000 just over five years later.

In a message published in OSHA's summer 2002 issue of *Job Safety & Health Quarterly*, the assistant secretary of Labor for OSHA, John Henshaw, said, "We estimate our 800-plus (VPP) sites cover over half a million workers and prevented more than 5,600 injuries in Fiscal Year 2000. As a result, we estimate these sites save more than $150 million annually in direct costs." Saving direct costs is another benefit realized by achieving excellence in an SH&E management system such as VPP.

Not only do businesses benefit from the internal results of the model SH&E program, but the business community is becoming a global enterprise rather than a local one. Industry is now worldwide. Safety, health, and environmental management systems must meet the expectations of many cultures. The Voluntary Protection Programs are broad-based in process and criteria—they easily fit the world-class category. They work in any culture.

This edition of the book is based on the revised criteria and the 1989 Guidelines. It covers the changes as part of the SH&E management system—from the first self-assessment to the first self-evaluation and to the continuing improvement. Readers can benefit from the techniques and methods in the first book along with adding others presented here.

Another vital element in this revised edition is the *reader*. This edition describes in more detail how all employees are involved in the management system—the executive "corporate conscience," the middle managers and supervisors, and the workers. Anyone who reads this edition will find in it something that he or she can do personally to enhance and support the company's SH&E management system.

To repeat what was said in the first book, "It's a win-win deal for employers, employees, and OSHA." Especially and most importantly for the employees anywhere in the world.

Charlotte A. Garner
Seabrook, Texas

Preface

What if there were safety program guidelines, available at no cost from the Federal government, that were proven to reduce injuries and improve your company's bottom line? There are—OSHA's Voluntary Protection Programs (VPP)! This book explains how to implement these guidelines—even if your company decides not to participate in the formal VPP Programs—and what to expect in terms of cost savings and reduced risk of injuries. As a Corporate Safety Administrator with Webb, Murray and Associates, Charlotte Garner has helped many organizations, including the Johnson Space Center achieve and maintain VPP status. As the author points out, the VPP program "…offers an organized, workable, time-tested set of guidelines that helps you to resolve the problems and challenges efficiently, proactively, and productively."

The tools for change Ms. Garner provides will allow SH&E managers to analyze their programs, build consensus and implement change. Ms. Garner's first ASSE book, co-authored with Patricia Horn, focused on benchmarking with OSHA VPP criteria. Mark Hansen, Past President of the American Society of Safety Engineers, wrote in a review of the book in *Professional Safety*, "**In my opinion books on safety and health need to be loaded with tools that safety professionals can take from print and use to enliven safety programs. This book is replete with such tools… .**"

How Smart Managers Create World-Class Safety, Health, and Environmental Programs provides SH&E managers with the set of principles based on OSHA VPP criteria, the tools to analyze program deficiencies, and proven consensus-building techniques that will result in effective implementation. Charlotte Garner convincingly explains why companies who have adopted these principles and practices have experienced improved operating efficiencies, reduced operating costs, and lowered incident and injury rates. Their new SH&E programs are founded on consensus and a shared vision, and provide companies with the flexibility to sustain continuous self-assessments, and quickly implement program improvements in order to remain competitive in a global economy.

1 Challenges and Opportunities to Prevent Injuries and Illnesses

If we keep doing what we have always done, we'll keep getting what we have always gotten. —*Anonymous*[1]

Will Americans eventually forget the horror, shock, and disbelief of the savage terrorist attacks on September 11, 2001, that murdered a total of 2,862[2] foreign and American citizens? Surely not. Since September 11, 2001, there is an urgency, a keener need to guard the safety and health of ourselves and our fellow humans. It permeates the home, the community, and the work site. We are more attentive to our own and to each other's welfare than we were on September 10, 2001. In the few minutes that it took for those four airplanes to slice through two enormous buildings, the Pentagon, and crash in Somerset County, Pennsylvania, our collective perspective on the worth of being human was transformed and transcended the past important materials in our lives.

From extolling the worth of having an excellent safety, health, and environment (SH&E) program for the sake of our workers, our collective attitude now easily accepts that it is essential to exert every effort to protect and add longevity to the lives of all. We finally realized that existence and survival are foremost.

On September 10, an excellent SH&E program was a "nice" thing to have. Why change what was already working? On September 11, an excellent SH&E program became "critical."

On September 10, many managers, safety professionals, and company executives thought their SH&E programs were "good enough to get by." Why spend the extra time, effort, and expense to attain so-called excellence?

On September 11, those same people realized their vulnerability to chance for the first time and "good enough" was not so good and "getting by" was not so comfortable. Excellence became instantly acceptable and absolutely necessary.

Let us trust that this sensitivity to the value of life is one positive outcome of the 9/11 tragedy, that it continues into the distant future, and that it is a constant reinforcement to meet and conquer not only the national and international security challenges but the challenges of implementing an excellent SH&E program.

What Is the State of Your Current SH&E Program?

In addition to the imperative to preserve life and health, there are other reasons to closely examine your current SH&E program. What is the state of that program? Is it comprehensive or piecemeal? Is it continuously improving? Is it reactive or proactive? Here are some questions to ask yourself:

1. Is top management actively and visibly involved in your SH&E program?
2. Do the employees participate in formulating work procedures and practices? That is, "Management demonstrates its commitment by 'setting an example by following the rules, wearing any required personal protective equipment, reporting hazards, reporting injuries and illnesses, and basically doing anything that they expect employees to do.'" (Extract from Chapter III, "Requirements for Star, Merit, Resident Contractor, Construction Industry, and Federal Agency Worksites," *Voluntary Protection Programs (VPP): Policies and Procedures Manual*, OSHA Directive Number TED 8.4, effective March 25, 2003.)
3. Do management, supervisors, and employees analyze the hazards of new or changed work before it is started?
4. Is OSHA and job-specific training regularly scheduled, administered, and documented? Does it include managers and supervisors?
5. Do you have a close call/first aid reporting procedure?
6. Do you and the employees involved investigate every incident, whether it is a close call or serious? Are corrective and preventive actions developed, assigned, and followed to completion?
7. Along with the line workers, are the supervisors and managers responsible and accountable for SH&E in their operations and departments? Is that reflected in their annual performance evaluations?
8. Do the managers and the supervisors regularly conduct scheduled site inspections and observations of work practices, and audit documentation and recordkeeping?

9. Do you have regular emergency evacuation drills?
10. Do you have an idea of what the direct and indirect costs are of injuries and illnesses at your company?

All of these questions and more refer to the elements that are part of an "excellent" SH&E program. These are a few of the items that OSHA compliance officers examine, inspect, or audit on a visit. They also run a tab on your injury/illness records to determine what the company ranking is compared to its industry average.

So—what *is* the state of your SH&E program?

Case Study

A company in disarray–the reactive SH&E program

The company that you work for may have the traditional safety and health department with one to several safety and health professionals to look after hundreds, or perhaps thousands, of employees. The company practice may be to comply only minimally with OSHA standards. There may be little interest in improving the quality of the safety and health program—until a mishap gets everyone's attention. For a transformation from "good enough" to "exemplary," a mishap does not have to be as tragic as a fatal injury to any employee. It can be serious, but not fatal; it can even be a major close call. But it will provide the necessary incentive to change the status quo!

At the moment of being notified of an accident—a fatality, an explosion, a fire, a chemical spill, a serious close call, or a mechanical failure—your daily routine will not only change instantly, but dramatically, and probably for the remainder of your career. Not only will you suddenly be enveloped by this tragedy; but new directions, new resolves, and new fears will immediately surface in your work-related tasks.

The investigations begin. There are meetings and meetings and meetings:

- Of department heads to discover the causes and the preventive measures. Reports are written, recommendations are made, and legal counsel is sought, when necessary;
- With a customer, if affected, including reports and explanations;
- With employees to restore trust and morale;

- To determine settlements with the injured or with the family of the deceased employee(s). There may be rehabilitation and a long period of recovery for the injured. Some of these issues may take years and BIG money to complete.

The scenario goes on; but you have the picture. The company worklife is consumed by this one accident for an unknown period of time. The more people involved, the more issues, the more complexities—the longer to resolution and closure.

For example, the author closely followed an industrial accident reported in the Houston, Texas, *Chronicle* in December 1996–February 1997[3], concerning an accident at a metal forging company in Harris County, Texas, that killed eight men. Five of the men were unscrewing a three-foot-wide, eight-inch-thick lid from a 90-foot nitrogen tank that was under 5000 pounds of pressure when the lid blew off, taking the workers with it. Three other men in the area were also killed by the blast and flying debris. Here are a few of the many issues resulting:

- The employer sent a company representative to personally notify the families.
- One OSHA inspector arrived at the site on December 23 and two more were en route. The team grew to five during the first days following the accident.
- A prayer service was held the day after the accident at the machinists' union hall; the plant was closed; no one worked that day.
- The company stock fell 1E/i points in heavy trading on Monday, December 23, the day after the accident.
- The victims' families hired attorneys.
- The employer deposited $10,000 for each family into the company credit union.
- One hundred workers were reassigned to other work; about 25 percent of the plant was affected by the explosion.
- As of January 4, 1997, debris and loose metal were still being cleared from the explosion site.
- The employer hired an engineering firm to investigate the accident.
- The employer donated $100,000 to each of the victims' families.

All of that and more, and still the day-to-day work had to continue. Only the plant owner knows how much were the total remedial costs relating to the accident. You who can calculate the direct and indirect costs related to an accident can estimate the total. Include in your estimate the $1.8 million in fines that OSHA assessed the company. They agreed to:

- Pay the fine within 12 days after the announcement;
- Make safety improvements in their five U.S. plants;
- Hire a full-time safety manager for the Houston operation;
- Retain an outside consultant to advise them and to certify to OSHA that procedures were being addressed;
- Strengthen the joint union-management safety committee.

This is the reactive safety and health program in action. This one safety and health problem disrupted and distracted the entire organization from top to bottom for months. This summary of the actions emphasizes the fact that safety, health, and environmental issues *are a constant, daily part of* every operation, function, and activity in the company. You cannot walk into any closet, office, or department or through any operation or activity of the company without finding SH&E issues, practices, and precautions. Issues that have such an intrusive, possibly extremely disruptive, potential to put the entire company into temporary disarray deserve the same close scrutiny, consideration, and oversight by the management and safety staff as all of their other top priorities, such as budgets, finances, people, dividends, products, quality assurance, and customer services.

Forewarned is forearmed. Did the preceding case study (one of thousands) serve as a wake-up call? Is yours a similar reactive SH&E program?

National Statistics: Fatalities, Injuries, and Illnesses

A combined challenge and opportunity exists in the data concerning work-related fatalities, injuries, and illnesses. The Bureau of Labor Statistics, Census of Fatal Occupational Injuries reports 5,524 fatal injuries in 2002.[4] The Bureau also reported 4.7 million nonfatal injuries and illnesses during 2002.[5] More information concerning numbers and costs is contained in Chapter 2. You will see that the annual costs and the number of days lost are staggering. This is a real, immediate, and life-saving challenge that affects every employer in the United States.

Any data related to a work-related death, injuries and illnesses, or to penalties and OSHA violations are data that should be considered by all owners, managers, and SH&E personnel as unnecessary, unacceptable, and unfortunate in the daily life of the company. Every effort should be willingly devoted to eliminate, reduce, and prevent the conditions that result in this data being part of the company's annual performance evaluation or planning. Our ultimate goal should always be no dead workers and continuous reduction in crippled and disabled workers.

Legal Requirement Challenges

Several requirements are already law and pose major SH&E-related legal challenges:

Americans with Disabilities Act (ADA)

The ADA ensures persons with disabilities equal opportunity for employment. Some specific positive steps must be taken by employers to accommodate the disabled as well as to ensure that compliance is being properly implemented.

The Department of Justice (DOJ) enforces the ADA's requirements in four areas:

Title I: Employment practices by units of state and local governments and by businesses with 15 or more employees.

Title II: Programs, services, and activities of state and local government.

Title III: Public accommodations and commercial facilities.

Title IV: Telephone companies must provide teletype (TTY) for the hearing impaired.

The DOJ use several methods to reach resolution of the enforcement activities:

1. Through lawsuits and both formal and informal settlement agreements. The DOJ is prohibited from filing a lawsuit unless it has first unsuccessfully attempted to settle the dispute through negotiations.

2. Through litigation, the DOJ may obtain compensatory damages and back pay as well as civil penalties up to $55,000 for the first violation and $110,000 for any subsequent violation.

3. In some cases, a negotiated consent decree from the violator. Consent decrees are monitored and enforced by the federal court in which they are entered.

4. The DOJ sometimes resolves cases without filing a lawsuit by means of formal written settlement agreements.

5. Informal settlements are reached frequently without litigation. In some instances, the public accommodation, commercial facility, or state or local government promptly agrees to take the necessary action to achieve

compliance. In others, extensive negotiations may be required but are concluded without litigation.

Knowledge of the ADA requirements can be helpful to an employer deciding when to hire an applicant who may have a pre-existing condition that might interfere with his/her performing the work without further injury. The OSH Act requires that employers shall protect workers from known or suspected hazards. On occasion, both OSHA and ADA should be researched for either conflicting or compatible requirements or implementation. On one of OSHA's Frequently Asked Questions web pages, OSHA was asked about an employee declining an audiometric test.

The OSHA response was:

> ...OSHA's Noise Standard requires only that audiometric testing be made available to all employees whose exposures equal or exceed an 8-hour time-weighted average of 85 dBA. On the other hand, the standard does not prohibit an employer from having a company rule that employees submit to audiometric testing. You should be aware, however, that the Americans With Disabilities Act (ADA) places certain limitations on employer-required medical examinations. Among other things, such examinations must be shown to be job-related and consistent with business necessity.[6]

ADA Regulations, technical assistance materials, and more information can be found on the ADA home page, Department of Justice website, www.usdoj.gov.[7]

OSHA Multi-Employer Site Citation Policy Directive CPL 2-0.124

This policy directive, issued December 10, 1999, states that on multi-employer worksites (in all industry sectors), more than one employer may be citable for a hazardous condition that violates OSHA standards. There is the creating, exposing, correcting, and/or controlling employer, each of whom may be subject to citation if found to be involved in the violation. The compliance safety and health officer can recommend certain abatement actions, and if the affected employer implements those actions, a citation can be averted.

This is a complex and confusing directive that must be studied carefully if you perform services for customers on their premises. If you have questions about it, you are encouraged to access the OSHA website (www.osha.gov) for the provisions of the directive and the administration of it by OSHA. Litigation appears not to have proliferated in this instance as it has for other controversial OSHA directives. To be on the conservative side, however, regard this directive as another set of rules that should be accommodated in your SH&E process with a contingency plan if you are in a situation where you could be cited under this rule. At the very least, become familiar with it.

29 CFR 1910.119, OSHA Process Safety Management of Highly Hazardous Chemicals Rule

29 CFR 1910.119 includes requirements for written safety and health procedures for all elements of the model SH&E program. If you are a small company, Appendix C of 1910.119 is informative regarding the scope and application of the rule to your SH&E program and your particular operations. The rule itself is mandatory; Appendix C is nonmandatory but it is recommended (by OSHA in the rule and by the author of this book) for the practical guidelines that are included.[8]

40 CFR Part 68, Environmental Protection Administration (EPA) Risk Management Program (RMP)

40 CFR Part 68 is a performance-based program that applies to some 15,000 U.S. facilities which possess sufficient quantities of hazardous toxic and flammable chemicals to present significant acute risk to the off-site public and environment. The facilities are required to periodically audit their management systems and correct any deficiencies. The requirements cover all of the elements that are in the model safety and health program being proposed by Congress (See "Codification of the model OSHA-approved SH&E" later in this chapter). Any source with more than a threshold quantity of a listed regulated substance in a single process must comply with the regulation. Compliance went into effect on June 20, 1999. The coverage of the RMP rule is not restricted only to large companies who process chemicals. It can also apply to smaller companies that may have only one process unit. Certain quantities of regulated substances determine compliance requirements. The quantities and regulated substances are defined in the rule. This is another rule that should be included in contingency planning if there is a potential for the company to be subject to compliance.

Construction Safety and Health Program

The Construction Safety and Health Program requirements can be found in 29 CFR 1926.20, General Safety and Health Provisions. Although not as comprehensive as the Voluntary Protection Programs, these requirements cover some of the basic elements of a model SH&E program. Under the Standards Interpretations Memorandum of August 22, 1994 (Revision 2, 9/20/95)[9], a construction site may be placed in a "focused inspection" category. Based upon the experienced judgment of the compliance officer, the construction company's SH&E program may be examined by the compliance officer.

Other Occupational Safety and Health (OSH) and Department of Transportation (DOT) Rules

Other OSH and DOT rules are also in place, and are extensive and comprehensive. It is incumbent on all employers to be aware of the requirements in these regulations that concern your operations. Not being aware is not an acceptable defense for any employer in this age of continuous, instant, and credible information and data available by several communication means. Electronically, OSHA and DOT regulations and quantities of information are available on the OSHA website, www.osha.gov, and the DOT website, www.dot.gov.

Environmental Compliance

Environmental compliance is another element that, in the last 10–15 years, has joined with the safety and health processes to prevent injuries and illnesses. Have you included the "environment" element? Have you considered the impact of environmental compliance on your operations? According to the EPA's published information, it appears that EPA has the authority to make painful inroads into the employer's pocketbook.

More about the EPA's specific costs is discussed in Chapter 2. For now, here is one bit of information from the EPA for your consideration:

> **EPA Achieves Significant Compliance and Enforcement Progress in 2002**
>
> The US Environmental Protection Agency has released data on its enforcement and compliance results for FY 2002, which included record-setting expenditures of almost $4 billion by violators for pollution controls and environmental cleanup. The program also secured commitments for an estimated reduction of more than 774 million pounds of harmful pollutants and the treatment and safe management of an estimated 2.8 billion gallons of contaminated groundwater. [10]

Employers should become alert to the environmental conditions inside their plants and operations. An environmental problem is another example of an unexpected, unplanned, reactive event that can seriously affect a company's survival.

National environmental compliance is not the primary focus of this book. The primary focus is on the safety and health of our workers in an environmentally clean work site. However, informing you of the magnitude of the EPA's national activities provides you with an understanding of the scope of their requirements. You can readily relate your budgets to the possible costs in your specific work environment.

Maintaining a pristine environment on our planet is of primary importance to our existence. It is incumbent upon all employers and employees to contribute their collective and individual efforts to comply with the EPA requirements or the environmental requirements that may exist in other parts of the world. This author is personally aware of instances when the customer required the contractor to submit an environmental control procedure as a part of pre-contract review. This is one reason that environmental activities are being included in the model SH&E program. Environmental awareness is a necessity, not a choice.

Ergonomics

Even though ergonomics is not a formalized OSHA rule at this time, the President and OSHA are using the general duty clause of the Occupational Safety and Health Act and various OSHA national emphasis programs (NEPs) to achieve compliance in specific targeted industries, such as the nursing home facilities.

> On April 5, 2002, OSHA announced a comprehensive plan to reduce ergonomic injuries through a combination of industry-targeted guidelines, tough enforcement measures, workplace outreach, advance research, and dedicated efforts to protect Hispanic and other immigrant workers.[11]

John Henshaw, the OSHA Administrator, stated that the agency would immediately begin developing industry and task-specific guidelines to reduce and prevent ergonomic injuries.

On the OSHA website you will find a page entitled, "Safety and Health Topics: Ergonomics," that discusses the "OSHA Effective Ergonomics Strategy to Success." Also, the ergonomics FAQ (Frequently Asked Questions) file linked to that page contains a detailed explanation of what OSHA ergonomics strategy involves.

Secretary Henshaw also stated:

> The Department's ergonomics enforcement plan will crack down on bad actors by coordinating inspections with a legal strategy designed for successful prosecution...For the first time OSHA will have an enforcement plan designed from the start to target prosecutable ergonomic violations. Also for the first time, inspections will be coordinated with a legal strategy developed by Department of

Labor (DOL) attorneys that is based on prior successful ergonomics cases and is designed to maximize successful prosecution.[12]

Along with these two key elements, OSHA will also offer Hispanic outreach, ergonomics research, and training. The secretary said that OSHA wants to work with the employers who are already voluntarily striving to reduce ergonomic risks without government mandates. Mr. Henshaw stated:

> We want to work with them to continuously improve workplace safety and health. We will go after the bad actors who refuse to take care of their workers.[13]

The concept of ergonomics touches every person who engages in work, recreation, home, and other activities (even warfare). Here also is another challenge and opportunity to voluntarily join in an OSHA cooperative effort to prevent and to reduce injuries and illnesses in the workplace.

Convergence of SH&E Programs

Another challenge and opportunity to reevaluate your current SH&E program lies in the similarity of several "excellent" SH&E programs shown in Table 1-1.

In December 1999, a group of students studying an internet-based class on occupational safety and health management, presented by a professor at Tulane University, were given a mid-term assignment to study the strengths and weaknesses of a model safety and health management program and compare it to the one being used at their places of employment. One of the students asked permission as a substitute assignment to report on the results from all of the programs selected by the students.

She made this request because she noticed that safety professionals and their managements had been devoting more attention to the elements and completeness of safety and health management programs. This heightened attention was due to three activities that were significant to the economic health of companies. One was the establishment by the International Organization for Standardization (ISO) of a quality assurance system (ISO 9001)[14] and an environmental management system model (ISO 14001). These two standards have been internationally accepted. Second is the introduction by OSHA of the Voluntary Protection Programs (VPP) to the industries of the United States, that are the program criteria presented in this book. Third, the organizations that have implemented the systematic safety and health program management process have realized significant improvements and savings.

The Tulane students surveyed twenty-eight mid- to large-size companies and universities. The results of the survey and subsequent report were published in an article by Peggy Farabaugh in the June 2000 edition of the *Occupational Health & Safety* magazine.[15] The students' summary of the occupational safety and health management systems is shown in Table 1-1.

Table 1-1. Characteristics of Occupational Health and Safety Management System Models

Model	Major Characteristics	Reference URL
OSHA VPP OSHA Voluntary Management Guidelines OSHA's proposed safety program standard	• Management leadership and employee participation • Hazard identification and assessment • Hazard prevention and control • Information and training • Evaluation	www.osha.gov www.vppa.org www.osha.gov/oshprogs/vpp/sitesalp.html www.osha-slc.gov/SSLTC/safetyhealth/inshp.html
AIHA-OHS Management System: a Guidance Document (American Industrial Hygiene Association AIHA 96/3/26)	• Systematic approach • Quality management • Ownership, accountability • Auditability • Integration with ISO 14001 and 9001	www.aiha.org click on "publications"
British Standards Institute BS8800	• Integration with overall management system • International quality management	www.bsi.org.uk
Chemical Manufacturers Association (CMA) Responsible Care (Author's note: CMA has been renamed the American Chemistry Council.)	Condition of membership in Codes of Management Practices: • Community awareness and emergency response • Pollution preventions • Distribution safety • Process safety • Employee health and safety • Product stewardship	www.cmahq.com/responsible-care.nst/pages/about
Joint Commission on Accreditation of Healthcare Organizations (JCAHO)	• Continuous improvement of safety and quality of care provided to public • Performance management	www.jcaho.org www.jcaho.org/perfineas/perfmeas_frm.html

Exhibit 1-1. (cont.)

Australian WorkCover Authority Safety Map	• Continuous improvement • Involvement of all workplace stakeholders in health and safety	www.workcover.vic.gov.au/vwa/projects.nst/all/SafetyMAP+New
Department of Energy (DOE) Integrated Safety Management System (ISMS)	**Core Functions** • Define scope of work • Analyze hazards • Develop and implement controls • Perform work • Feedback and improvement **Guiding Principles** • Line management responsibility for safety • Clear roles and responsibilities • Competence commensurate with responsibilities • Balanced priorities • Identification of safety standards and requirements • Hazard controls tailored to work being performed • Operations authorizations	http://tis.eh.doe.gov/ism/ http://tis.eh.doe.gov/ism/course.index.htm
Det Norske Veritas OHSMS and International Safety Rating System	• Focus on management activities rather than just compliance issues • Continuous improvement • Employee involvement	www.dnv.com

The organizations that have issued these similar, but different, SH&E programs have concentrated on what can be done to, as the article states:

> ...prevent injuries and illnesses and the associated human and economic costs, not to merely abide by the law. It is interesting to note another similarity. None of the models is currently "required" by governmental regulation. If OSHA succeeds in ushering a proposed model safety and health rule (the VPP) into a regulation, it will be the first OHSMS (Occupational Health and Safety Management System) required by law.[16]

It behooves us to seek the challenges and opportunities for life-saving techniques, systems, and processes in the *prevention* of work-related injuries and illnesses. Such opportunities exist in these similar SH&E processes and systems that are being initiated, encouraged, and implemented by so many governmental and global organizations.

Different industries, different organizations, different governments, different countries. Even though each one has different requirements, there is a convergence of elements in each that is common to all. Each one of these elements involves the practices that have proven effective over the years to result in a quality product or service, or the completion of the necessary tasks with order, efficiency, and competency. The SH&E program is a new factor in the equation. Only recently, in the last 25–30 years, have SH&E factors become an equal and sometimes even a dominating factor in the work environment along with the product, marketing, services, and finances. Efficiency, quality, and accuracy, not SH&E, have been the primary company focus. Discipline and enforcement have been the guidelines to create a safe work environment. They have not succeeded.

The organizations and regulatory bodies in Table 1-1 have recognized that adversarial enforcement of rules is not accomplishing the intent of law. Where SH&E is concerned, work practices are being addressed, not adversarial disputes. Work practices concern people, machines, complex systems, work conditions, the environment, and other elements, and require cooperative communication to be effective. Certain responsibilities rest with employers and employees to perform the work efficiently, safely, accurately, and productively within the scope of their responsibilities and within a culture of cooperation and communication.

Why should you take note of these activities? It is a trend that has developed over the last fifteen years and is expanding globally. It appears that the excellent SH&E program, although an option now, may be a mandatory requirement in the not-too-distant future.

Codification of the Model OSHA-Approved SH&E Program

During the initiation and implementation of the programs in Table 1-1, the various entities recognized that the excellent programs contain the same essential elements. Apparently, if this trend continues, the SH&E program with common characteristics will eventually be a requirement, not an option. This is another challenge/opportunity on the horizon.

On April 15, 1999, "The Models of Safety and Health Excellence Act of 1999, (H.R. 1459) was introduced in the House of Representatives by Congressmen Thomas Petri (R. WI) and Robert Andrews (D. NJ). The bill, which would codify the OSHA Voluntary Protection Programs (VPP), has experienced bipartisan support and is now on its way to becoming law."[17] In fact, in the Winter 2002 issue of *The Leader*, 12 Democratic congressmen, 27 Republican congressmen, and one Independent congressman are identified as members of the strong bipartisan group that are supporting passage of this bill.[18] The Secretary of Labor has also stated that President George W. Bush supports the bill and will approve it when it is passed.

The idea of voluntary protection has progressed in popularity since 1999 when the Models Act was introduced. Logically it follows that federal codification formally recognizing voluntary compliance is likely to be a reality in the foreseeable future. It will become the standard by which all other SH&E programs are evaluated. See Appendix A for the text of the Models Act.

There are advantages and benefits both in accident prevention and improved economic health that encourage an employer to advance to the front of the line in voluntarily adopting the model excellent safety, health, and environment program. The advantages of this type of proactive SH&E program will be discussed throughout the book.

Your Relationship with OSHA

Compliance with OSHA standards is mandatory. Since you must comply, it is smart to implement the best program available, one that will accrue benefits for your workers *and* the company. You are encouraged to implement all of the criteria to achieve an effective, proactive SH&E program as well as one that will pass OSHA inspection and that will prepare you for contingencies in the future.

A significant advantage also is that if excellent program criteria are the foundation of your SH&E program, you will find the compliance officers in a much friendlier mood when, or if, they come to see you. The current OSHA compliance philosophy is to cooperate and work with companies that are striving for excellence in their SH&E program and to penalize the "bad actors" to the full extent of the law. This is another advantage to voluntarily implementing a model SH&E process.

Summary

In this chapter, you have been introduced to the major SH&E challenges and opportunities with which the SH&E people, managers, and top managements are confronted in the present volatile economic environment. The greatest present challenges relate to legal requirements from several governmental sources, the Occupational Safety and Health Administration, the Department of Justice, the Department of Transportation, and the Environmental Protection Agency. Also included is the ergonomics compliance program that the Department of Labor plans to mount through the enforcement door of the OSH Act.

The characteristics of a reactive SH&E program and an example case study of a company whose SH&E program is in disarray are presented. An emerging trend of national and international entities who have identified the elements of an excellent SH&E program is also described. The commonalties of these programs are identified.

The proposed codification of a model safety and health program by the U.S. Congress is included as another indicator of the trend toward mandating the excellent safety and health programs. The employer's relationship with OSHA is presented. It can be adversarial or cooperative. The era of enforcement only is past. The current governmental attitude toward SH&E is to recognize and reward the employers who establish and maintain exemplary SH&E programs and vigorously penalize the serious violators who, with informed negligence, still put worker lives at fatal risk.

Chapter 2 addresses the costs of not having an exemplary SH&E program, including the costs of OSHA penalties, the number of OSHA inspections conducted, and the leading violations that OSHA penalizes. Other costs are also discussed, including workers' compensation and costs relating to injuries, illnesses, and accidents.

No matter what the challenges are and what opportunities are available to improve your SH&E system, be assured that there *is* a management system that will contribute to the internal control you can exert over these challenges. The system can favorably influence the well-being of your workers and the economic health of the company, no matter what external pressures are placed on it. It's a win-win deal.

Recall the opening quotation: "If we keep doing what we have always done, we will continue to get what we have always gotten." Maybe it's time for a change in how our SH&E processes are functioning.

References

1. Anonymous, 2001.
2. Bureau of Labor Statistics, *Census of Fatal Occupational Injuries Summary*, September 25, 2002.
3. *The Houston Chronicle*, 12/24/96–2/12/97.
4. Bureau of Labor Statistics, *Census of Fatal Occupational Injuries Summary*, September 25, 2002.

5. Ibid. *Workplace Injuries and Illnesses 2002*, December 18, 2003.
6. OSHA website, *OSHA FAQs* (various topics), [www.osha.gov], accessed August 10, 2003.
7. Americans with Disabilities Act, ADA Business Connection website, [www.usdoj.gov/crt/ada/business.htm], accessed April 14, 2002.
8. OSHA Regulations website, [www.osha.gov], accessed April 2002.
9. Stanley, James W. "Guidance to Compliance Officers for Focused Inspections in the Construction Industry." *OSHA Standards Interpretations*, August 22, 1994 (Revision 2, 9/20/95).
10. US Environmental Protection Agency EPA Newsroom website, [www.epa.gov/newsroom/], accessed February 3, 2002.
11. US Department of Labor, Office of Public Affairs. "OSHA Announces Comprehensive Plan to Reduce Ergonomic Injuries" (USDL 01-201). News release: April 5, 2002.
12. Ibid.
13. Ibid.
14. Garner, Charlotte, and Patricia Horn. *How Smart Managers Improve Their Safety and Health Systems: Benchmarking with OSHA VPP Criteria*. Des Plaines, IL: American Society of Safety Engineers, 1999: 12–13.
15. Farabaugh, Peggy. "OHS Management Systems: A Survey." *Occupational Health & Safety Magazine,* June 2000: 44.
16. Ibid: 42.
17. "VPP codification Bill Introduced," *The Leader*, Summer 1999.
18. Ibid, Winter 2002: 18.

2 Why Do This?
The Bottom Line

The evidence of VPP's success is impressive. Recent data show VPP worksite injuries and illnesses that keep employees away from work or necessitate their restricted work activity or job transfer are dramatically below industry experience. As a result, VPP worksites have saved more than a BILLION dollars since the program began in 1982. In addition, many VPP participants report workplace improvements such as lower turnover rates, reduced absenteeism, and improved employee morale.[1]

Executive Questions

"What is this remodeling going to cost? What are the downsides? Why should I want to do this? What's in it for the company and the stakeholders?" These may well be some of the questions that company owners will ask themselves or you. Perhaps the question that they should also be asking is, "How much does SH&E cost the company *not* to work toward excellence in these processes?"

Implementing quality criteria and attaining excellence is hard work, time consuming, does not happen overnight, and can possibly cost more money and effort than is possible for the company to dedicate at one time. This book, however, will point to ways and means to get the program implemented effectively while taking into consideration the expenditure of monetary and human resources. The implementation of a model SH&E program will, in the long term, result in greater rewards than the expenditure of the effort, time, and money during the implementation. As with any endeavor, dedication and perseverance are integral ingredients for excellence in this system.

According to the data they have released, the companies that have implemented a model SH&E management system have found the rewards more than compensate for the short-term intensive labor and resources required. OSHA states in *OSHA Facts* (from the OSHA website, May 19, 2002) that the model SH&E program "continues to pay big dividends" to the companies that have implemented it.

Why Do This?

In response to the question, "How has VPP improved worker safety and health?" OSHA answered on the OSHA VPP website (in a release dated September 1, 2003):

> Statistical evidence for VPP's success is impressive. The average VPP worksite has a lost workday incidence rate 52 percent below the average for its industry. These sites typically do not start out with such low rates. Reductions in injuries and illnesses begin when the site commits to the VPP approach to safety and health management and the challenging VPP application process.

External Pressures on the Bottom Line

OSHA penalties

As we all are keenly aware, compliance with OSHA rules is mandatory. Most of the standards cost something to implement, and some of them cost quite a lot. Other than the cost to implement, there are the costs of *not* complying. These costs should be considered the most wasteful of company revenues. Whatever your level of management in a company, you are not working as hard as you do to establish a successful business and show a healthy return on investment just to see part of company profits disappear down the noncompliance black hole.

OSHA's federal and state inspection revenue in 2002 was slightly over $149 million for 95,000-plus inspections. Was any part of your company's profit dollars included in that $149 million? The costs of noncompliance, whether high or low, are costs that can and should be eliminated from a company's ledgers.

Also, consider that OSHA will assess the highest penalty allowed for noncompliance in the coming years. The rationale is that there is no acceptable excuse for noncompliance. Knowledge of requirements is readily available. Employers of any size in even the most remote locales of the United States know there are SH&E rules that must be followed, just as they know that they must be licensed for conducting their businesses and they must pay taxes on the revenues.

In early 2003, OSHA published their Strategic Management Plan for FY2004–2008:

> The OSHA areas of emphasis will be analyzed and revised each year based on the result of operations and new issues that demand attention.

Those emphasized areas are construction first, general industry second, and ten other categories. Some of the others are "high industry/high severity industries, oil and gas field services, concrete and concrete products, amputations in manufacturing and construction" plus six others.

Regardless of the industry and VPP participation, OSHA will still respond to serious accidents, fatalities, chemical spills, and worker complaints. OSHA statistics indicate that inspections are by no means limited to the targeted 3,000 worksites and major incidents. Note that state OSHA inspections (58,402) exceeded the number of federal OSHA inspections (37,493) in 2002.[2]

In addition, OSHA announced that, during fiscal year 2002, the number of trained, certified OSHA inspectors would be increased "to make its enforcement efforts more effective."[3] Strong, well-funded OSHA compliance activities will continue for the foreseeable future. Enlightened employers, union stewards, and SH&E professionals should heed these storm warning flags and take whatever steps are necessary to attain and maintain an OSHA inspection-free SH&E management system.

EPA compliance costs

As mentioned in Chapter 1, the Environmental Protection Agency (EPA) is also making major inroads into the pocketbooks of any industry that has contaminated, or has the potential to contaminate, the environment. From the EPA website, "EPA National News," accessed on January 30, 2003, EPA announced that its "strategic and smart enforcement and compliance assurances show results in cleaner air, water, and land."[4] The enforcement and compliance assurances in fiscal year 2002 mean that almost $4 billion will go towards cleaning up polluted sites and protecting the environment. Other EPA enforcement and compliance monetary returns and actions are shown in Table 2-1.

As noted in Table 2-1, EPA enforcers are a good deal tougher on environmental violators than OSHA is on SH&E violators. OSHA does not often cite an employer in a civil or criminal court, much less have the violators incarcerated for months or years. A jail sentence and a high dollar fine are two primary means of EPA's attaining compliance effectiveness.

Why Do This?

Table 2-1. Highlights of EPA FY 2001–2002 Enforcement Penalties and Compliance Activities

	FY2001	FY2002
Commitments by violators	$4+ billion	"Almost" $4 billion
Administrative, criminal, and civil judicial penalties	$148+ million	$144 million
Compliance inspections	~5,000	17,600
Willful violators' sentences	256 years	215 years

Sources: US Environmental Protection Agency, "EPA National News," 1/02/02 and 1/20/03.

Be proactive. If you have an environmental compliance plan, review it at least annually to ensure that it is up to date and at the appropriate level of readiness. If you do not have an environmental compliance plan, put one in place as soon as you can—especially in this uncertain era of American businesses and citizens being threatened by terrorism. Be at the ready!

Internal Pressures on the Bottom Line

Workers' Compensation and other costs

The National Safety Council states in the 2003 edition of *Injury Facts:*

> According to the National Academy of Social Insurance, an estimated $49.4 billion, including benefits under deductible provisions, was paid out under workers' compensation in 2001 (the latest year for which data were available), an increase of about 3.5 percent from 2000. Of this total,
>
> - $27.4 billion was for income benefits and
> - $22 billion was for medical and hospitalization costs.
>
> In 2001, approximately 127 million workers were covered by workers' compensation—a decrease of 0.1 percent over the 127.1 million in 2000.

There are other costs relating to injuries and illnesses that must be considered. According to the National Safety Council *Injury Facts,* 2003 edition, there were:

- 3.9 million disabling injuries in a workforce of 136.2 million in 2001;

- 3.7 million disabling injuries in a workforce of 137.7 million in 2002.

Tables 2-2, 2-3, and 2-4 display some of the costs incurred relating to injuries and illnesses.

Table 2-2. Estimated Work Injury Costs, 2001–2002

Nature of Losses	2001 (Billions)	2002 (Billions)
Wages and productivity	$69.2	$74.0
Medical	$24.6	$27.7
Administrative expenses	$21.7	$26.3
Includes:		
Money value of time lost by others directly or indirectly involved to conduct and record investigations	$11.8	$12.5
Damage to motor vehicles	$2.0	$2.8
Fire losses	$2.8	$3.3
Total Work Injury Costs	$132.1	$146.6

Source: National Safety Council, *Injury Facts*, 2002–2003.

Table 2-3. Estimated Costs per Worker

Cost	2001	2002
Per worker	$970	$1,060
Per fatality	$1,020,000	$1,070,000
Per disabling injury	$29,000	$33,000

Source: National Safety Council, *Injury Facts*, 2002–2003.

Table 2-4. Days Lost Due to Work Injuries

Cause	Days Lost, 2001 (Millions)	Days Lost, 2002 (Millions)
Injuries (not day of injury or follow-up treatment)	85	80
Injuries in prior years	45	45
Total days lost	130	125

Table 2-4. Days Lost Due to Work Injuries (cont.)

Cause	Days Lost, 2001 (Millions)	Days Lost, 2002 (Millions)
Estimated total days lost in future years from injuries occurring in 2001–2002	65	65

Source: National Safety Council, *Injury Facts*, 2002–2003.

The NSC goes on to state, "Not included is time lost by persons with non-disabling injuries or other persons directly or indirectly involved in the incidents." In an article published by the American Industrial Hygiene Association,[5] Robert A. Brennecke presented a detailed list of the direct and indirect costs. Even as estimates, in many of the accidents the injuries and the costs represent preventable pain and suffering by the workers and unnecessary losses and expenditures by the employers. These enormous costs and the penalty costs do not mean that more workers are being hurt. In fact, the trend of injuries and illnesses is downward. It means that the costs are rising.

OSHA states in *OSHA Facts:*

> In 2001, occupational injury and illness rates dropped to the lowest level—5.7 injuries per 100 workers—since the U. S. began collecting this information, part of an eight-year downward trend. The total number of worker deaths in 2001 was 5,900.[6]

Although these figures indicate a downward trend in fatalities, even *one* work-related death is one too many.

Any increase in accident-related costs is also unacceptable. There seems to be no feasible quick fix to reduce the costs. The best recourse for employers at this time is to initiate an energetic and workable SH&E process to reduce injuries and illnesses with a continuous goal of prevention.

Brennecke's article provides an insightful analysis of how you can estimate the effect of safety and health incidents upon your profit. An example of Brennecke's formula for correlating costs to profits concerns your premium for workers' compensation insurance. If the same special trade contractor cited in the article pays the usual premium rate of $46,000 per year, the contractor would have to make $1,150,000 in sales just to pay the premium. If the contractor's premium were $20,000, sales would only have to be $500,000. The lower your premium, the fewer sales are required to offset the cost. The better your safety and health program and the fewer incidents experienced, the lower your premium, and the greater the company's profit.

You can download OSHA's "$AFETY PAYS"[7] software program from the OSHA website (www.osha.gov). This program calculates worker's compensation costs and the product sales required to meet those costs. The formula used by the program is the same one used by Brennecke and Fulwiler (see below) to calculate costs.

Richard D. Fulwiler, director, health and safety for Proctor & Gamble at that time, stated the following in a 1991 paper that is relevant even a decade later:[8]

> ...The contribution that health and safety makes to the company's profits may not always be dramatic or highly visible, but it is there just the same. For example, worker's compensation costs are one valid measure of the effectiveness of an organization's health and safety program.
>
> To demonstrate this point, let's look at Procter & Gamble's 1989 worker's compensation experience. Our workers' compensation costs were about $10 million. Comparative data show that these costs are from four to eight times better than the national average. Using a mid-range of 6, Procter & Gamble's worker's compensation costs for the same period would have been about $60 million. Further, by using our 1988 earnings figure of 5.6 percent, the company would have had to increase sales by over *$1 billion* just to offset these increased workers' compensation costs...Thus, it can be seen that being "average" in health and safety carries an extremely high price tag. A corporation with an "average" health and safety performance simply will not survive in today's highly competitive environment...

Few companies are the size of Procter & Gamble at the time Fulwiler cited his case. The figures cited are dramatic simply because of the amounts. No matter the size of the company, however, these examples illustrate the severe effect that safety and health program performance can have on profits.

A well-known teacher, author, and expert in the field of construction safety, Jimmie Hinze, states:

> The true costs of worker injuries are often not fully appreciated. The insurance premiums become a routine expenditure, and the indirect costs of injuries are generally absorbed in other line items in the budget. That is not a wise accounting approach, especially if those costs are not known. Closer examination of total costs of injuries reveals that they are often quite

high. Managers...should become fully aware of the magnitude of those costs. Managers who have examined such costs in greater detail have generally responded by placing greater emphasis on the safety aspects of...work....The costs of injuries are high, even when they do not include the pain and suffering of the workers and their families. Once managers fully understand the magnitude of the true costs of injuries, it is easier for them to adopt a zero-accident philosophy.[9]

Uneasy, cynical, distrustful workforce

Another significant internal factor may influence you to implement the OSHA-approved SH&E program criteria. "Though Upbeat on the Economy, People Still Fear for Their Jobs," reads an article headline in *The New York Times* of December 29, 1996. Downsizing and restructuring and outsourcing and other reorganization strategies in the last decade have nearly destroyed worker confidence that today's jobs will be there for them tomorrow. The *Times* article says:

> Through November of this year (1996), even after more than five years of economic recovery, the number of layoffs announced by companies had increased 14 percent from a year earlier...Seventy percent of those [workers] polled said they thought layoffs were not just a temporary problem in America but a feature of the modern economy that would continue permanently. Last year (1995), 72 percent said layoffs would continue permanently...After all, more than half of American workers are employed by companies with 500 workers or fewer..."People worry about their own jobs, not jobs as a whole," said Eric Greenberg, director of management studies for the American Management Association. "In the current environment, a sense of job insecurity on the part of the individual is very rational."[10]

It seems safe to assume that the environment of worker uncertainty and insecurity exists to this day.

If job insecurity was a reality in the upbeat economy of 1996, consider its magnitude in today's economy. Since many people have the perception that they cannot depend on their employers for long-term employment, they also have the perception that they have the right to change jobs for better pay, better opportunity, or better benefits. Loyalty to an employer is hard to come by these days. After all, if their employers are not going to be loyal to them, why should they be loyal to their employer? No matter the size or business of your company, it may have gone through some reorganizing or restructuring during the last five to ten years. Only your top managers know if job insecurity is or is not present among the employees as a result of reorganizing.

Even in the work environment of the second millennium, basic uneasiness is still present in worker attitude. The focus has shifted somewhat from disbelief in the permanency of a job to a search for better opportunities.

Gregory P. Smith, in a more recent article for Business Know-How (www.businessknowhow.com)[11] stated:

> Today's workplace is different, diverse, and constantly changing. The typical employer/employee relationship of old has been turned upside down. The combination of almost limitless job opportunities and less reward for employee loyalty has created an environment where the business needs its employees more than the employees need the business.... Management's new challenge is to transform a high-turnover culture to a high-retention culture...

On May 14, 2002, HRMGuide.net's headline was, "Job-changing rate still high in 2001."[12] The article cited a survey of 1,477 workers conducted for CareerBuilder (www.careerbuilder.com) that indicated:

> ...job-changers focus on salary, location, and work-life balance. The survey showed that almost four out of ten U.S. workers were looking for new opportunities in 2001. Although they were generally satisfied with their jobs, more workers were considering changing jobs in 2001 than in 2000. Diane Strahan, career specialist with CareerBuilder, called them "the silent job seekers who are constantly on the lookout for new opportunities and who remain optimistic, even with the recent news of layoffs and restructurings...Don't confuse their silence and satisfaction with job complacency...

Whatever the conditions in your industry and in your company, the current "temporary job" mindset of the general work population generates indifferent behavior and attitudes toward the job and encourages frequent absenteeism and turnover. Frequent turnover and lack of job-specific skills are major factors that contribute to increased work accidents, serious injuries, mechanical failures, production delays, waste, and other adverse operational results. Rebuilding or maintaining mutual trust and trustworthiness with employees is a worthy goal for any employer in today's volatile economic environment.

Why Do This?

Strong Internal Control

To maintain smooth and uninterrupted product output, there are indicators within company operations that are watched closely, such as the peaks and valleys of cash flow, poor delivery of raw materials, and delayed shipments of products to customers. At such times, management takes immediate remedial action. At other times, unplanned or unexpected events—such as natural disasters, fires, job-related death or serious injury to one or more employees—can cause major disruptions. If SH&E management does not ensure that contingency plans are developed to handle such occurrences, everyone in the company is confronted with scores of clean-up problems along with maintaining regular operations.

With strong internal controls in place, the impact of temporary setbacks can be minimized. Exerting control over or planning for events that can cause loss of money, people, or other resources is one of the major responsibilities of SH&E and other managerial leadership. As the numbers cited in these chapters signify, maintaining vigilant personal control within each individual's area of responsibility for safety and health management systems can significantly contribute to a reduction of injuries, illnesses, absenteeism, turnover, and waste; and an increase in profitability.

Evidence of the reality possible through strong control and dedicated administration of a company's SH&E systems comes from some of the model SH&E program participants. "In the three years we worked to qualify for the Star (VPP) Program, we reduced our lost workday cases by 74 percent and our workers' compensation costs by 88 percent," says Phillip B. Chandler, safety and environmental superintendent, Thrall Car Company, Winder, Georgia.[13]

In 1994, OSHA cited a Monsanto spokesman:

> Monsanto Chemical Company's Gonzales, Florida, plant experienced a steady decline in its lost workday case rates during the period the worksite was implementing effective safety and health programs and in the four years since approval to the VPP. The rates fell from 2.7 in 1986 to 0.15 in 1993.[14]

A Mobil Oil spokesman as cited by OSHA:

> Mobil Oil Company's Joliet, Illinois, refinery experienced a reduction in its lost workday case rate from 3.8 in 1987, the year before it began implementing VPP quality safety and health programs, to 0.1 in 1993, two years after approval to the Star Program. In the same period, the refinery experienced a drop of 89 percent in its workers' compensation costs.[15]

Motorola, Incorporated states on its website:

> Eight of our sites have earned the U. S. Occupational Safety and Health Administration's (OSHA's) highest award—the Voluntary Protection Program (VPP) Safety Through Accountability and Recognition (STAR) award...This prestigious national award is issued to sites demonstrating superior safety and health programs...For the past five years, Motorola's global injury and illness rate has been significantly below both the U. S. manufacturing average and the U. S. electronics industry average. In 2001, we met our goal of continuously improving our injury and illness case rate over previous years and reduced our rate to 0.96 recordable injuries and illnesses per 100 employees.

In 1997, Motorola's global recordable injury and illness rate was slightly over 6.0 per 100 employees.[16]

These large companies and some 1,000 others are now profiting by implementing a systematic, proactive SH&E process. The process is certainly worth the attention of SH&E management and company leadership to learn more about this credible system to conserve worker health and welfare, reduce outlay, and increase income.

Future Considerations

Other factors, not known or not clear at present, also threaten a company's competitive edge, increasing the need for company-wide stability and durability. Technological advancements in production, manufacturing, information, and services, and availability (or lack) of resources, raw materials, and competent people must be included in the company's long-term planning. Regulatory requirements, current and evolving, must be considered as a major factor in plans. Experts can forecast certain economic developments under various conditions, *but no one can know with certainty*. Internal control based on firmly fixed organizational principles and systems will help the leadership steer a steady course through the unknown future with an educated and prepared certainty of reliable and favorable outcomes.

What Will It Cost?

One answer is that to implement a model SH&E system of programs will surely *not* cost as much as the cost of *not* implementing the model management system in terms of work-related deaths, injuries, and illnesses and the cost of *not* complying with the regulations. But in reality, what the cost will be to bring the SH&E system up to excellence can only be determined by measuring the current program against the criteria for the model program, identifying the gaps and omissions that must be closed, and estimating the costs in time and resources to close them.

The gaps and omissions may or may not be inexpensive. Some may require future planning and budgeting. Or they may be repaired in the short term. There are solutions and options, however, to accommodate whatever it takes in the budget and planning cycles to meet the model system's criteria, that will be acceptable to OSHA, and that will not consume resources that may not be available in quantity at a given time. The next chapters will explore these questions and offer workable solutions to achieve excellence in SH&E with a healthy workforce and without unnecessary reduction of the bottom line. In Chapter 3, the self-assessment yardstick is described to assist in measuring the current SH&E process and identifying the gaps.

The Problem and the Solution

One of today's major challenges for SH&E professionals, company leadership, and employees, is to formulate a philosophy and vision for a durable, stable company that operates on principles and values of trustworthy management commitment, frank and truthful openness, meaningful employee involvement, and mutual trust, and that has the resilience to respond to external demands with dependable, proactive, innovative, creative, and positive systems and processes. The solution: start now. Evaluate your organization for these characteristics. Where there are weaknesses, start building strengths. Where there are strengths, make them stronger.

Start with this project that has already been tested and measured. Work through it. Assess, evaluate, and measure the company's SH&E process. Benchmark the process against participating leaders in your company's industry.

The quotations cited from specific companies that have implemented model SH&E management systems illustrate that this is a win-win strategy for workers, management, and OSHA or whatever international jurisdiction under which the employer operates. The OSHA Voluntary Protection Programs managers in local and regional offices can refer you to companies in your own industry that will serve as mentors and advisors. The VPP Participants Association (See Appendix B, *Sources of Help*) will also refer you to contacts for information.

The VPP SH&E management model is an ongoing process of continuing improvement wherever improvement is needed. The result, and one solution to the challenges present and foreseen, is a vitalized, upbeat, productive SH&E management and worker team functioning with proven management principles and processes—an organization prepared and positioned for whatever the future brings.

Summary

This chapter discusses the external pressures threatening the company's bottom line, such as costly OSHA penalties, new and pending OSHA rules that measure or will measure the effectiveness of a company's safety and health program, and future considerations.

Internal corporate considerations are presented, such as the staggering costs of worker deaths, injuries, and illnesses, and the temporary-job mindset of today's workforce. The increasing cost of injuries and illnesses compared to the decreasing incidence of injuries and illnesses indicates that the two do not correlate. More importantly, these costs and penalties emphasize the pain and suffering—sometimes death—of the American worker and the hardships placed upon their families because of the worker's disability, either temporary or permanent. The chapter suggests that the staggering financial costs and penalties that employers are experiencing may be due to the need for stronger internal control of their volatile SH&E processes.

Credible sources cite estimates of what the injury and illness cost equivalents are in corporate revenue and pre-tax corporate profits. Examples of companies that have implemented the model SH&E criteria indicate reduced injury and illness costs and increased profitability.

Also discussed is the demoralization of the workforce attributed to downsizing, restructuring, and outsourcing and the related high turnover and downturns in productivity and sales. What the cost may be to implement the VPP requirements, how long it will take, and the improved position of the company after implementation are addressed.

One possible and long-lasting solution to the problem of profitably doing business is offered—a solution that will endure into the future, foreseeable or unknown—the VPP SH&E management model.

Chapter 3 discusses conducting an assessment of your current program to determine where it stands in comparison to the model system's criteria.

References

1. Occupational Safety and Health Administration, Region VI Management Office. "VPP—Voluntary Protection Programs—Recognizing Excellence in Safety and Health." Dallas, Texas, August 2002.

2. 2003–2008 Strategic Management Plan, Occupational Safety & Health Administration website, [www.osha.gov], accessed 1/9/04.
3. US Department of Labor, Office of Public Affairs, *OSHA Facts*, 2003.
4. US Environmental Protection Agency, "EPA National News" website, [www.epa.gov], accessed 1/30/04. "EPA Achieves Significant Compliance and Enforcement Progress in FY2003."
5. Brennecke, Robert A. "Safety and Health vs. Profit: Contributing to the Corporate Bottom Line." *The Synergist,* August 15, 1997: 27–28.
6. US Department of Labor, Office of Public Affairs, *OSHA Facts*, 2003.
7. US Department of Labor, "$AFETY PAYS" software, available on the Occupational Safety & Health Administration website [www.osha.gov], accessed 9/1/03.
8. Fulwiler, Richard D., Director, Health and Safety, Procter & Gamble Company, from a paper prepared for the Minerva Education Institute, Xavier University, Cincinnati, OH, 1991: 2.
9. Hinze, Jimmie W. *Construction Safety.* Upper Saddle River, NJ: Prentice Hall, 1997.
10. "Though Upbeat on Economy, People Fear for Jobs." *The New York Times,* Sunday, December 29, 1996, "National" sec, p. 15.
11. Business Know-How website, [www.businessknowhow.com], accessed May 14, 2002.
12. HRMGuide.net website, [www.hrmguide.net], "Job-changing rate still high in 2001," accessed May 14, 2002.
13. US Department of Labor, Occupational Safety and Health Administration, "Safety and Health is Good Business—OSHA Voluntary Protection Programs."
14. Ibid.
15. Ibid.
16. Motorola Occupational Health and Safety website, [www.motorola.com/EHS/safety/] accessed June 1, 2003.

3

Is This Model SH&E Program for You?

SH&E Management System Self-Assessment

...The most difficult and most important decisions in respect to objectives are not what to do. They are, first, what to abandon as no longer worthwhile and, second, what to give priority to and what to concentrate on.... These are...and should be informed judgments. Yet they should be based on a definition of alternatives rather than on opinion and emotion. The decision about what to abandon is by far the most important and the most neglected.... An organization, whatever its objectives, must therefore be able to get rid of yesterday's tasks and thus to free its energies and resources for new and more productive tasks.

—*Peter F. Drucker*[1]

Change Is Ever with Us

As Drucker pointed out in 1969, change is ever with us. Change will always be the wave of the future. "Opportunities for improvement" (a.k.a. changes) are with us permanently.

There are periods in everyone's experience, both as companies and as people, when we have to make difficult decisions without being certain of a beneficial and productive outcome. These decisions may have concerned financial affairs, downsizing, a major change in employment or plant operations, or advanced technology that appeared to offer opportunities for growth.

Before a company makes those kinds of difficult decisions, management must first carefully assess the depth and scope of the effect of change upon all of the company's operations. The next step is to decide what are the specific remedies, corrections, or eliminations that management will need to make to move the company onto a more productive and profitable track.

After planning the renewal effort, executives, senior staff, managers, and employees embark on the corrected course. This is the same process used in the model SH&E management system: reviewing the outcomes of current processes company-wide, deciding what, if any, remedies are needed, planning the route through the forest of corrections and adjustments, and moving forward to excellence. The question for you is whether the present company SH&E practices are relevant to the present and the foreseeable future or whether they should be modified, or perhaps even abandoned, to optimize the economic worth of the company and the welfare of the employees.

Keep in mind that, sometimes, radical measures are necessary to accelerate the company's performance and to keep up with competitors who are offering a wider range of options. Markets and competition are worldwide and brutal. Sophisticated competitors and customers are joined with rapid change. These factors exert greater and greater pressures on suppliers to offer quality products at lower prices.

Change and innovation are accelerating with intensity. With these volatile conditions, companies are compelled to examine carefully any and all obsolete procedures that are no longer effectively accomplishing their purposes—those that are not preventing, reducing, or eliminating injuries or illnesses, and that are not cost effective. Some form of redesign or reengineering must rebuild or replace these broken systems. Companies must have effective systems that add to, not decrease, the company's strengths and that allow continuous updating and improvement. Speed, innovation, flexibility, quality, service, and cost are critical to the survival of today's organization.

It is a good idea to benchmark the model SH&E programs of other companies in your company's industry (even your keenest competitors). Look at the cost of injuries and illnesses and of noncompliance and the negative impact of these two nonproductive activities on the economic health of your company. It becomes apparent that the SH&E program and the company will benefit from the ongoing surveillance inherent in and required by the SH&E model management system.

How does the SH&E model management system differ from traditional safety and health programs?

There are requirements in the VPP SH&E model management system that have not been so heavily emphasized in traditional programs. Foremost is that management and employees own and are actively involved in implementing and sustaining the management system. There is no more "safety is not my job" attitude. The model SH&E system intentionally involves *everyone* in the company—top management, middle managers, line supervisors, and all employees. It is a structured, organized, systematic management process that addresses the risks of the workplace logically and thoroughly. In Table 1-1, note that in the eight management systems, there is a similarity of characteristics, functions, and operations.

Not only does the model management system help you to identify the problems and challenges existing in the company's current SH&E process, but it offers an organized, workable, time-tested set of guidelines that helps you to resolve the problems and challenges efficiently, proactively, and productively. The primary purpose of this book is to offer this credible and doable process that, once woven into the fabric of the company's culture, will not only identify the deficiencies but also offer the flexibility to accommodate fitting solutions over the long term.

Maybe you have thought that this approach to SH&E management does not compare in depth and scope to the other issues that affect company operations. Not so. From the first two chapters, it is apparent that SH&E has joined with other significant issues that the company must take into account to survive. It is important that the company make every effort to improve this particular system, since the first two chapters show that a dominant factor in the company's profitability is the SH&E management system, which is or can be among the most cost-effective systems in the company. It contributes noticeably to company profitability and to the protection of the employees—two indispensable management responsibilities.

Is this SH&E management system for you?

To help you answer this question, take a closer look at this model SH&E management system. Understand that by adopting this program, the company is in no way obligated to formally participate in OSHA's approval process. It is a strictly voluntary effort.

Here are some excerpts about the Voluntary Protection Programs (there are three levels of programs, Star, Merit, and Demonstration) taken from OSHA literature.[2] More detail will be added as we progress through the chapters.

> ...Working with industry and labor, OSHA created the VPP in 1982 to recognize and partner with worksites that implement exemplary systems to manage worker safety and health. The managers, employees, and any authorized representatives at these sites voluntarily implement comprehensive safety and health programs—hereinafter referred to as safety and health management systems—that go beyond basic compliance with OSHA standards...
>
> Using one set of flexible, performance-based criteria, the VPP process emphasizes holding managers accountable for worker safety and health, the continual identification and elimination of hazards, and the active involvement of employees in their

own protection. These criteria work for the full range of industries, union and non-union, and for employers large and small, private and public...

The VPP place significant reliance on the cooperation and trust inherent in partnership. Sites choosing to apply for VPP recognition show their commitment to effective worker protection by inviting a government regulator into their workplace. In return, OSHA removes them from programmed inspection lists and does not issue them citations for standards violations that are promptly corrected...

Sites qualifying for VPP attain Star, Merit, or Demonstration status. Star participants meet all VPP requirements. Merit participants have demonstrated the potential and willingness to achieve Star status, but some aspects of their programs need improvement. Demonstration participants test alternative ways to achieve safety and health excellence that may lead to changes in VPP criteria...

Statistical evidence for VPP's success is impressive. Consistently over its 20-year history, the average VPP worksite has had an incidence rate for days away from work, restricted work activity, and/or job transfer that is at least 50 percent below the average for its industry!...

OSHA VPP reviewers don't look for a single correct way to meet VPP requirements. They want to see a system that works for you. Some successful safety and health management systems involve substantial written documentation, and others do not. Small businesses, in particular, often are able to implement excellent safety and health processes with relatively little documentation...

Stephen Brown, Union Safety Representative, PACE Local #712, Potlatch Consumer Products Division, Lewiston, ID, is quoted: "My site first began researching the VPP in 1994. In 1995, we started pursuing our goal to become an OSHA Star site seriously, and in 1996 we were awarded Merit status. That year we broke all records. We have repeated that every year since then, and I am proud to say we attained Star status in 1998."...

...The VPP process of managing safety and health systematically—using one set of flexible, performance-based criteria—is working in workplaces large and small. VPP has been effective in various industries—from construction to poultry processing to petrochemical plants, from tree nurseries to nursing homes, and from mom-and-pop operations to federal laboratories. It works in union and non-union shops. More than 180 distinct industrial classifications are represented, and the number is growing. VPP has proven more successful in reducing work-related injuries and illnesses than mere compliance with specific regulations...

"For those of you who have not yet joined the ranks of VPP, I'd like to take this opportunity to encourage you to do so. Not often do you get a chance like this to forge a new relationship with a regulatory agency based on trust and cooperation. This is a powerful partnership with all long-lasting results. It's also a tremendous responsibility, yet a smart one, that will pay your business back," from Rich Guimond, Vice-President and Corporate Director of Environment, Health and Safety, Motorola, Schaumburg, IL.

First Action: Get Top Management Approval

Of course, nothing is done in a company without top management approval. The first action is to present to management the data in Chapters 1 and 2 along with the model SH&E criteria. These data demonstrate how the program will improve the quality of the SH&E management system and, also, improve the company's competitive edge and profitability.

OSHA recommends that a preliminary self-assessment of the current program be conducted to determine if the company's SH&E would qualify to apply for VPP approval. A self-assessment provides a baseline toward improvement. Include a recommendation to conduct the self-assessment when you make the presentation to management for their approval.

After obtaining approval to proceed, for a large operation the next step is to organize a steering committee composed of major company stakeholders to examine the critical elements *in detail.* Some of the major stakeholders are the company's human resources, production, quality control, marketing, maintenance, financial, and key union and non-union employees. In a small operation, determine what help, if any, you need to conduct the assessment. Maybe only the SH&E responsible person and one or two managers, supervisors, and employees can conduct the assessment. Or, select the committee members from other significant or unique components of the company's organization.

If possible, not only should a member of management or a designated management representative chair the steering committee, but that person should also chair the subcommittee that examines the management commitment requirements. That subcommittee should be empowered to assess candidly and with immunity the status and degree of management's actual ownership commitment of the process and identify the weaknesses and strengths of that commitment. Critical to the success of this SH&E management system is that management participation in the process be dealt with as strictly as is expected by any one of the subcommittees evaluating other elements.

Second Action: Conduct an Assessment of the Present SH&E Program

A reputable, credible set of criteria should be used to measure the company's current SH&E management system. In order to know what needs to be fixed, you first have to find out what is broken. To find what is broken requires a standard of expected performance.

The VPP are external criteria that have proven worthy of implementation and are used as a measuring rod of excellence in 982 companies[3] in many industries (as of July 31, 2003). The criteria can be confidently used as benchmarks to compare to the company's SH&E management processes. This is the step when you compare the company's SH&E processes to the VPP SH&E management system, collect the company's data and information, and assess the quality and effectiveness of the current programs.

Four critical elements of the model SH&E management system

From OSHA's perspective[4] there are four critical elements that must be part of the company's SH&E management system:

- Management leadership and employee involvement;
- Worksite analysis;
- Hazard prevention and control;
- Safety and health training.

Management leadership and employee involvement

The two complementary aspects of this element are the most critical to the success of any company-wide venture, especially in the VPP management system. They are the two aspects that only corporate management can initiate and move forward.

Enlightened management leadership provides the motivating force and the resources for organizing and controlling activities within an organization. In

Second Action: Conduct an Assessment of the Present SH&E Program

the VPP process, leaders regard worker safety and health as a fundamental and valuable activity of the organization and apply energy and attention to safety and health protection with as much vigor as to the other organizational purposes and goals.

Employee involvement provides the means through which workers develop and express their own commitment to safety and health protection for themselves and for their fellow workers. Employees effectively participating in the SH&E management system become stakeholders. They then develop a sense of ownership for the success of the venture and support it.

Here are the functions included in OSHA's self-assessment checklist as a part of management commitment and employee involvement:

__ **A managerial commitment to worker safety and health protection**

Managers must personally and actively perform their SH&E responsibilities to such a degree that all employees recognize and accept that management's personal commitment to the goal of excellence in the SH&E system is just as dedicated as its commitment to the profitability of the company. OSHA says:

> ...Actions speak louder than words. If top management gives high priority to safety and health protection *in practice*, others will see and follow. If not, a written or spoken policy of high priority for safety and health will have little credibility, and others will not follow it.

> Plant managers who wear required personal protective equipment in work areas, perform periodic 'housekeeping' inspections, and personally track performance in safety and health protection demonstrate such involvement...[5]

During this assessment, leadership can discover if their involvement has sufficient force to influence the plant superintendents, supervisors, and foremen to:

- Wear their required personal protective equipment.
- Submit regular reports of housekeeping inspections, documenting what was found and how and when it was corrected.
- Submit reports of the safety performance in their areas with an analysis concerning trends of reductions or increases of close calls, first aid, or more serious incidents and recommendations for appropriate corrective action.

If these activities are not happening, once again leadership must question what is wrong—whether it is disbelief in leadership's sincerity, an insufficient visibility of leaders, or leadership's inconsistent and unconvincing enforcement of the new guidelines—and what corrections must be made.

The material that follows is adapted from *Revisions to the Voluntary Protection Programs to Provide Safe and Healthful Working Conditions*.[6] Appendix C is a verbatim extract of Section III, "The Voluntary Protection Programs."

___ **Top site management's personal involvement**

OSHA says:

> Each applicant must be able to demonstrate top-level management leadership in the site's safety and health program... Managers must provide visible leadership in implementing the program. This must include:
>
> • Establishing clear lines of communication with employees;
>
> • Setting an example of safe and healthful behavior;
>
> • Creating an environment that allows for reasonable employee access to top site management.

___ **A system that is in place to address safety and health issues/concerns during overall management planning/purchasing/contracting**

Management systems for corporate-wide planning must address protection of worker safety and health. In Appendix D of the OSHA VPP evaluation guidelines[7], one of the points that must be covered in the assessment/evaluation is how, or if, safety and health practices are integrated into comprehensive management planning. That is, are expenditures for safety and health included in the annual budget planning? Scheduling the annual SH&E self-evaluation to coincide with the annual corporate planning schedule will allow you to budget for the SH&E objectives along with other company initiatives for the coming year.

In this assessment, you are expected to review your written safety and health program to discover if it contains all of the critical elements, including budgeting safety and health needs.

___ **Safety and health management that is integrated with your general day-to-day management system**

Authority and responsibility for employee safety and health must be integrated with the overall management system of the organization and must involve employees.

Second Action: Conduct an Assessment of the Present SH&E Program

___ **A written safety and health management system—often referred to as a safety and health manual—with policy and procedures specific to your site, appropriate for your site's size and your industry that addresses all the elements in this checklist**

All critical elements of a basic SH&E management system must be a part of the written program. As previously stated, those are management leadership and employee involvement, worksite analysis, hazard prevention and control, and safety and health training.

___ **A safety and health policy communicated to and understood by employees**

If the company management has not issued a written SH&E policy, signed and dated by the owner/executive of the company, and distributed company-wide, the leadership assessment subcommittee should recommend that this be done before any other action with the support and approval of the owner/executive. OSHA says,

> …your top level safety policy specific to your facility. Management must clearly demonstrate commitment to meeting and maintaining the requirements of the VPP.

The policy emphasizes the value that management places on SH&E requirements and purpose. If this value is internalized by all of the employees in the company, it becomes the basic point of reference for all decisions affecting SH&E. It also becomes the criterion by which the adequacy of protective actions is measured.

The assessment questions should search for validation that the policy is so well stated that all employees not only understand the priority of SH&E in relation to other organizational values, but they also are living it day to day and nothing interferes with that first priority.

___ **Safety and health management system goals and results-oriented objectives for meeting those goals**

The goal should be broadcast throughout the company along with the objectives to attain the goal. Although there may be an underlying goal stated in the SH&E management document that addresses zero accidents, injuries, and illnesses, the goal for assessment purposes is the annual specific goal or goals to accomplish for the coming year. This could be something like, "Our goal is to reduce the number of incidents in the manufacturing unit and introduce the employees to the use of job safety analyses (JSAs) as a training method." Along with the goal, develop the objectives to achieve the goal; such as, "This year, we will develop job safety analyses for the jobs in the manufacturing unit and administer JSA training to the unit employees."

Then, when you assess the effectiveness and the results for the manufacturing unit's training, you look at the number of incidents occurring this year compared to previous years. In many companies, there may be different goals with related objectives for different operations. It depends on how many are considered essential relative to the resources that are available. If you are a small or medium-size company, you can have as few as you feel that you can support for the year. Other goals are placed on the "To Do Later" list.

Look for evidence that the goal and the objectives are sufficiently specific to direct the affected employees to the desired results and to the methods for achieving them. Communicating the goal and objectives points the employees in the right direction to excellence. The subcommittee wants to know if the employees *are* going in the right direction, are they taking the correct actions, and have those actions resulted in the outcomes that are anticipated. If not, why not? Even if the answer is yes, can the system be further improved?

It may be necessary to reconsider the goal and objectives, to ask critical questions about the decisions and directions, to adjust the forward path, and to modify the goal and the objectives where the assessment indicates incompatibilities.

___**Clearly assigned safety and health responsibilities with documentation of authority and accountability from top management to line supervisors to site employees**

Ownership is established by investing in the venture, whether it is buying a home, starting a business, or working at a job. The same scenario applies to establishing ownership in the SH&E processes. In the prior version of the VPP Guidelines, OSHA noted that when responsibility for safety and health protection is assigned to a single staff member or even to a small group, other staff members and employees develop the attitude that someone else is taking care of safety and health problems—"that isn't my problem." *Everyone*—all levels of employees—has some responsibility for safety and health in the model SH&E management system. Each person at each level has his or her individual and functional safety and health-related activities; and each person at each level should know what those are. Not only should they *know what* their specific duties are, it is essential that everyone *understands* and is trained in *how* to perform them.

The assessment of this area should reveal that all employees know what is expected of them. If they do not, question them and the system to dis-

Second Action: Conduct an Assessment of the Present SH&E Program

cover why not and initiate corrective steps. If they are responding as planned, then leadership has the pleasure of commending them for a job well done and asking for suggestions to improve the system.

Ensure, also, that the responsibilities for specific SH&E functions are assigned, that the authority and resource provisions are specified and are known, and that they are drawn upon by all levels of employees to meet their responsibility expectations.

A declaration that certain performances are expected from managers, supervisors, and employees means little if management is not serious enough to track performance, to reward it when it is well done, and to correct it when it is not.

It is standard operating procedure that management holds everyone who works for the company accountable for meeting their job responsibilities. That is the essence of any effective and profitable company. Holding employees accountable for their safety and health responsibilities and duties is another integral component of the company's plans for success and stability.

The system of accountability must be applied to everyone from senior and corporate management to hourly employees. If some are held firmly to expected performance and others are not, the system will lose its credibility. Those held to the expectations will be resentful and those allowed to neglect expectations may continue to downgrade the quality of their work. The result is that the chances of serious injury and illness can increase.

Leadership must personally ensure that the accountability system is fair, reasonable, and consistently applied, no matter who the offender may be or what status in the company he or she may have. Check the system closely during the assessment to verify that such accountability is in place and that the rules are impartially administered, no matter who is the offender.

__ **Necessary resources to meet responsibilities, including access to certified safety and health professionals, other licensed health care professionals, and other experts, as needed.**

It is counterproductive to assign responsibility without providing commensurate authority and resources to get the job done. To do so sets the employee up for frustration and disappointment and sets the SH&E management system up for failure. A person with responsibility for the safe operation of a piece of machinery should have the authority to shut it down and get it repaired when it malfunctions.

Needed resources include operational and capital expenditure of funds as well as responsible, well-trained, and equipped personnel. The revised SH&E management system documents assignment of authority. In a later chapter, the strategic plan includes provision for resources.

Professional industrial hygienists, health care, and safety personnel must also be available when their services are needed.

___Selection and oversight of contractors to ensure effective safety and health protection for all workers at the site

The VPP include the requirement that the employer's SH&E management system must include the provision that the contract and construction workers are ensured safe and healthful working conditions consistent with and equal to those of company employees, and that their safety performance is tracked by the same system and procedures that are applied to the company employees.

The contractors must follow worksite safety and health rules and procedures applicable to their activities while at the site. The worksite employer is expected to encourage the contractors to develop and operate an effective safety and health program management system.

The worksite employer must have in place a documented oversight and management system for applicable contractors. "Applicable contractors" in the revised VPP Guidelines are defined as "those employers who have contracted with the site to perform certain jobs and whose employees worked a total of 1,000 or more hours in at least one calendar quarter at the worksite."[8] The documented oversight and management system must ensure that the contractors' site employees are provided effective protection that drives improvement in contractor safety and health. Such a system should ensure that safety and health considerations are addressed during the contractor selection process and when contractors are onsite.

___At least three ways employees are meaningfully involved in activities and decision making that impact their safety and health

Since an effective program depends on commitment by employees as well as managers, it is important that the systems reflect their concerns. As mentioned earlier, to be effective the VPP management system must include all personnel in the organization—managers, supervisors, *and* employees.

By this all-inclusive requirement, OSHA in no way intends to transfer responsibility for compliance to employees. The OSH Act clearly places responsibility for safety and health protection on the employer. In order to attain the optimum SH&E performance, however, OSHA has found that

Second Action: Conduct an Assessment of the Present SH&E Program

everyone in the company must be involved. Important to attaining the best performance is the employees' intimate knowledge of the jobs they perform and the special concerns they bring to the job. This gives them a unique perspective that can be used to make the program more effective.

Employees should have an impact on the decision process through methods such as hazard assessment, inspections, safety and health training, and/or evaluation of the SH&E management system. These methods are discussed in more detail in Chapter 8.

During assessment, the subcommittee wants to know if those methods are available to the employees; if a work environment has been created that welcomes, encourages, and supports employee participation; and whether it is happening. The subcommittee also wants to learn if the participative activities are resulting in improved safety and health program performance. Are there verification and validation procedures in place to measure the activities and to yield an analysis of the results?

__ **Annual safety and health management system evaluations on VPP elements in a narrative format, recommendations for improvements, and documented follow-up.**

The results of this self-assessment establish a baseline from which the SH&E planners can discover the weaknesses and the strengths of the company programs. Under the VPP requirements, a self-evaluation should be conducted annually. It is a tool to assist in ensuring the success and stability of the company's SH&E management system.

In OSHA's words:

> …the comprehensive program audit evaluates the whole set of safety and health management means, methods, and processes, to ensure that they are adequate to protect against the potential hazards at the specific worksite.[9]

This is the one VPP program element that OSHA has found generally lacking or misunderstood by sites applying for VPP. That is unfortunate, because when the annual evaluation of the management system is consistently administered, all of the processes are constantly improving. More will be discussed in the chapter on measuring the SH&E model management system from year to year.

__ **Formal signed statements from all collective bargaining agents indi-**

cating support of your application to VPP

___Where no collective bargaining agent is authorized, written assurance by management that employees understand and support VPP participation

The above two provisions are in the VPP requirements for the company that may apply for formal acceptance. Whether or not your employer plans to apply for VPP acceptance, these two requirements are included here to alert owner/management that they should ensure that all union and non-union employees are informed of what the model SH&E management system is, how it functions, what results are expected, and how vital their participation is to the success of the system. Also, these requirements contribute to the ultimate and never-ending goal of reducing or eliminating work-related minor, serious, or fatal injuries and illnesses. It is a matter of all-inclusive communication with everyone in the company.

Worksite analysis

Hazard analysis of routine jobs, tasks, and processes should be conducted regularly to identify uncontrolled hazards and lead to hazard elimination or control.

A *baseline hazard analysis* identifies and documents common hazards associated with your site, such as those found in OSHA regulations, building codes, and other recognized industry standards, and for which existing controls are well established.

Hazard analysis should also be conducted on significant changes, including non-routine tasks, new processes, materials, equipment, and facilities, so that uncontrolled hazards are identified prior to the activity or use. This will lead to hazard elimination or control. Important considerations include:

- **Samples, tests, and analyses.** Samples, tests, and analyses to identify health hazards and employee exposure levels that follow nationally recognized procedures must be conducted regularly and as needed.

   Documentation of the sampling strategy is necessary to identify health hazards and accurately assess employees' exposure, including duration, route, frequency of exposure, and number of exposed employees.

- **Self-inspections.** Self-inspections shall be regularly conducted at least quarterly (weekly for construction work) by trained staff, with written documentation and hazard correction tracking. These self-inspections shall cover the entire site.

- **Written hazard reporting system.** A written hazard reporting system that enables employees to report their observations or concerns to management without fear of reprisal and to receive timely responses is crucial.

- **Accident/incident investigations.** Sites should have a written procedure for investigating accidents, near-misses, first-aid cases, and other incidents. The procedure should include investigators' training, and what incidents warrant investigation. It should also state that accident/incident investigations are conducted by trained staff with written findings that aim to identify all contributing factors, corrective actions required, tracking actions to complete, and a summation of what action was taken to prevent similar events in the future.

- **System to identify and analyze incident data.** A system should be in place that analyzes injury, illness, and related data, including inspection results, observations, near-miss and incident reporting, first aid, and injury and illness records, to identify common causes and corrections in systems, equipment, or programs.

Hazard prevention and control

An effective system for eliminating or controlling hazards must be in place. This system emphasizes engineering solutions to provide the most reliable and effective protection. It may also use, in preferred order:

1. Administrative controls that limit daily exposure, such as job rotation.
2. Work practice controls, such as rules and work practices that govern how a job is done safely and healthfully.
3. Personal protective equipment.

All affected employees must understand and follow the system. Key elements include:

- **Tracking hazard correction.** A system for tracking hazard correction should be operational. It should include documentation of how and when hazards are identified, controlled, or eliminated, and communicated to employees.

- **Written preventive maintenance system.** A written preventive/predictive maintenance system that reduces safety-critical equipment failures and schedules routine maintenance and monitoring.

- **Occupational health care program.** An occupational health care program appropriate for your workplace. It should include, at a minimum, nearby medical services, staff trained in first aid and CPR, and hazard analysis by licensed health care professionals, as needed.

- **Consistent disciplinary system.** A consistent disciplinary system must be enforced and applied to all employees—including supervisors, managers, and contractors—who disregard the rules.
- **Written emergency response program.** Written plans to cover emergency situations, including emergency and evacuation drills for all shifts, should be available to all employees.

Safety and health training

Administer and document training for:

- Managers and supervisors, emphasizing safety and health leadership responsibilities;
- All employees on the site's safety and health management system, emphasizing hazards, hazard controls in place, and the VPP;
- All employees to recognize hazardous conditions and to understand safe work procedures applicable to their work environments;
- Assessing employee comprehension and training effectiveness.

A review of the SH&E documentation that is maintained by the company should be included in the self-assessment. The VPP criteria expect the documentation in Exhibit 3-1 to be part of the SH&E management system. It should be tailored to the business of the company and its particular SH&E activities within each critical element.

Third Action: Identify the Gaps

Now that you know what you will look for in the self-assessment, how do you get started and what do you do with it once it is finished? The VPP evaluation guidelines in Appendix D contain the details for conducting the assessment. Use the guidelines to learn the details of the parts that should be included in the company's current safety and health program and to discover:

- What is working well, and what isn't;
- What contributes to accomplishing the goal and objectives, and what doesn't;
- What is being done because it has always been done, but doesn't contribute to the mission, goal, or objectives of the program or the company;
- What is missing, and what is incomplete.

During the assessment, interviewing the employees and the site walk-through is not necessary. These two functions are part of the self-evaluation that will be conducted annually. For the purpose of the preliminary

Third Action: Identify the Gaps

Exhibit 3-1. Documentation Required to Meet VPP Criteria

(Excerpt from *Revisions to the Voluntary Protection Programs to Provide Safe and Healthful Working Conditions*, Federal Register 45646-456633, 7/24/2000)

Onsite document review will entail examination of the following records (or samples) if they exist and are relevant to the application or to the safety and health program (trade secret concerns will be accommodated to the extent feasible):

1. Written safety and health program
2. Management statement of commitment to safety and health
3. The OSHA Form 200/300 log for the site and all site contractor employees who are required to report
4. Safety and health manual(s)
5. Safety rules, emergency procedures, and examples of safe work procedures
6. The system for enforcing safety rules
7. Reports from employees of safety and health problems and documentation of management's response
8. Self-inspection procedures, reports, and correction tracking
9. Accident investigation reports and analyses
10. Safety and health committee minutes
11. Employee orientation and safety training programs and attendance records
12. Base safety and industrial hygiene exposure assessments and updates
13. Industrial hygiene monitoring records, results, exposure calculations, analyses and summary reports
14. Annual safety and health program evaluations, site audits, and, when needed to demonstrate the VPP criteria are being met, corporate audits that a site voluntarily chooses to provide in support of its application. The review of evaluation documents needed to establish that the site is meeting VPP requirements will cover at least the last 3 years and will include records of follow-up activities stemming from program evaluation recommendations.
15. Preventive maintenance program and records
16. Accountability and responsibility documentation, e.g., performance standards and appraisals
17. Contractor safety and health program(s)
18. Occupational health care programs and records
19. Available resources devoted to safety and health
20. Hazard and process analyses
21. Process Safety Management (PSM) documentation, if applicable
22. Employee involvement activities
23. Other records that provide relevant documentation of VPP qualification

NOTE: With regard to this documentation and to small business participation in VPP, OSHA states (Section IIIF5a(4) of the above document): "All aspects of the safety and health program must be appropriate to the size of the worksite and the type of industry. For small businesses, OSHA may waive some formal requirements, such as certain written procedures or documentation, where the effectiveness of the systems has been evaluated and verified. Waivers will be decided on a case-by-case basis."

assessment, examination of programs, processes, and documentation is sufficient to identify the deficient or weak areas in the current program.

The scope and depth of the critical elements are contained in Appendix E, "Questions that OSHA VPP Evaluators Ask When Conducting Pre-Approval Site Evaluations." It provides more information about what OSHA expects each critical element to contain.

In this volatile and swiftly changing economic environment, nationally and internationally, this is a good time to instigate a keen inward look at the managers' and employees' deepest, most personal values. Examine closely their willingness to change where their values are in conflict with new processes, new technology, new functions, new marketing, and expanding, more competitive markets—and the negative impact that the current SH&E systems are having on the welfare and health of the employees and the bottom line of the company.

Summary

Chapter 3 has outlined the areas of the VPP criteria that OSHA considers to be the vital elements and functions of an excellent SH&E system. The self-assessment should compare the company's current SH&E system to these criteria. As stated earlier, you have to know what is broken before you can fix it. This chapter identifies what elements and functions your SH&E system needs to produce outstanding performance in the reduction of injuries, illnesses, and environmental problems. Where the program is lacking and what actions can be taken to fill the gaps are discussed in the following chapters.

Now is the time to conduct an in-depth gap analysis to determine what is lacking or deficient and how it can best be resolved. Gap analysis is covered in Chapter 4. Chapter 5 discusses the strategic plan that can be used to address the gap resolutions with an orderly approach that considers who, what, when, where, and how of closing the gaps.

References

1. Drucker, Peter F. *The Age of Discontinuity: Guidelines to our changing society.* New York: Harper & Row, 1969: 192.
2. OSHA Region VI website, [www.osha.gov/dcsp/alliances/regional/region_6.html], "VPP: Voluntary Protection Programs, Recognizing Excellence in Safety and Health," accessed July 2002.
3. OSHA, Office of Partnership & Recognition, Department of Labor, Washington, DC, July 31, 2003.
4. Ibid.

5. OSHA Preamble of *The Voluntary Protection Programs Management Guidelines,* Federal Register, 1989.
6. Occupational Safety and Health Administration. *Revisions to the Voluntary Protection Programs to Provide Safe and Healthful Working Conditions,* Washington, DC: Federal Register 65-45649-45663, July 20, 2000: 20–27.
7. Ibid.
8. Ibid.
9. Ibid

4 Where Do You Want to Be?

Gap Analysis

...Gap analysis is an active process of examining how large a leap must be taken from the current state to the desired state—an estimate of how big the "gap" is. The "analysis" provides the answer to the question of whether the skills and resources at hand are sufficient to close the gap—to achieve the desired future within the proposed period.
—*Goodstein, Nolan and Pfeiffer*[1]

Now you must make those "most difficult and most important decisions...first, what to abandon as no longer worthwhile and, second, what to give priority to and what to concentrate on..." (Drucker, *The Age of Discontinuity*).[2] Other important considerations and decisions await; such as, which programs or processes are deficient, which do not exist, how much effort will it take to fill the gaps, who will be affected, and what, where, and how severe will be the repercussions.

Building the Bridge Between Now and the Future

The self-assessment has indicated the current state of the company's SH&E program. Incidents for the past one to three years have been identified, along with their severity, their frequency, what caused them, and the corrective actions needed to prevent recurrence. Management systems in place to prevent or eliminate the causes before the incidents occur have been examined.

Now, look at where the SH&E system would be if the VPP criteria were in place. The guiding questions in making these decisions are:

- How much better would the program be if this particular gap were closed?
- Is this program necessary or just nice to have?
- What is the cost in time, money, effort, and other resources?
- How many of these resources are available? Can they be used for this activity?

- How long will it take?

All of the actions cannot be accomplished at the same time; but you can plan and schedule to get them all completed over time. OSHA does not expect all actions to be performed at once. This is the continuous improvement process required by the VPP. The OSHA VPP criteria state *what* essential elements should be established to achieve an excellent SH&E management system performance and *what* the performance results should be. They do not, however, prescribe the *how* to implement the criteria to get from "establish essential elements" to "performance results." OSHA expects and prefers that the "how" be tailored to the functions of the company.

Here also, you consider the necessity to rid yourself of the old processes and establish entirely new ones. There could be gaps on which not much action is required. But if it is a gap, it is because something is required that is not in place or not up to criteria and it should not be lightly passed over. Some current processes can be eliminated because they are so obsolete that they are no longer useful or are being overtaken by newer processes.

> Executives, whether they like it or not, are forever bailing out the past…Today is always the result of actions and decisions taken yesterday…Yesterday's actions and decisions, no matter how courageous or wise they may have been, inevitably become today's problems, crises, and stupidities. Yet it is the executive's specific job…to commit today's resources to the future…But one can at least try to limit one's servitude to the past by cutting out those inherited activities and tasks that have ceased to promise results…
> —Peter Drucker, *The Effective Executive*[3]

Once you have reviewed the findings of the self-assessment, your major challenge is pruning the "inherited activities and tasks" that no longer contribute positive results or have become obsolete. You know what activities and tasks are being considered and you know how eliminating them will affect the people who are performing them.

These are tough decisions according to Peter Drucker.[4] A few pointers on effective decision-making may help. The rationale for decisions, as suggested here, will also be useful in explaining to those affected why the decision was made. It may be that the affected employees participate in putting the new systems and processes in place, with training in the acceptance and management of change. Get them involved. It will make the decision easier to execute.

Where does environmental compliance fit?

Even though environmental compliance is not a part of the VPP criteria, many companies are including it as another function of the department with safety and health responsibilities. The recommended course for you is to include environmental requirements as part of the self-assessment and treat it as you would the other gaps.

It is also a management system and process. References are included in Appendix F to provide environmental requirements against which your program can be measured or developed.

Characteristics of Effective Decision-Making

Refer to Exhibit 4-1 for a flow chart that illustrates the characteristics of effective decision-making.

1. **Leadership must establish a rule or a principle that is fundamental, unchanging, and general from which other rules are derived for long-term solutions.**

Since the VPP criteria will affect all of the company operations, decisions relating to their implementation require overriding principles.

One of the fundamental definitions of a process, a rule, or a principle is that it is a foundation, a benchmark against which the quality, completeness, and rightness of operational results are determined. Drucker says to ask yourself, "Is this something that underlies a great many occurrences?"[2] The answer is "yes" to the systems and processes that apply the VPP criteria to managing the multitude of safety and health conditions, situations, and occurrences that arise daily, monthly, or yearly in the course of a company's life. The principle or rule resulting from the decision must provide a guiding sense of the requirements and obligations of right conduct and management.

2. **The decision must be far reaching and provide an umbrella for all other implementing rules.**

Define the basic parts of the problem that the decision process must address to solve this and other problems.

The decision must include all of the related conditions so that it results in the resolution of the problems over the long term. The leadership of the SH&E change should insist that advisers and the steering committee present a concise, clear description of the complete situation as it is; not as they want it to be. An example is the required comprehensive baseline survey of hazardous conditions. Inform the group that a complete report is expected on the state of the current program, as are recommendations for a

Exhibit 4-1. Flow Diagram of Decision-Making Steps[6]

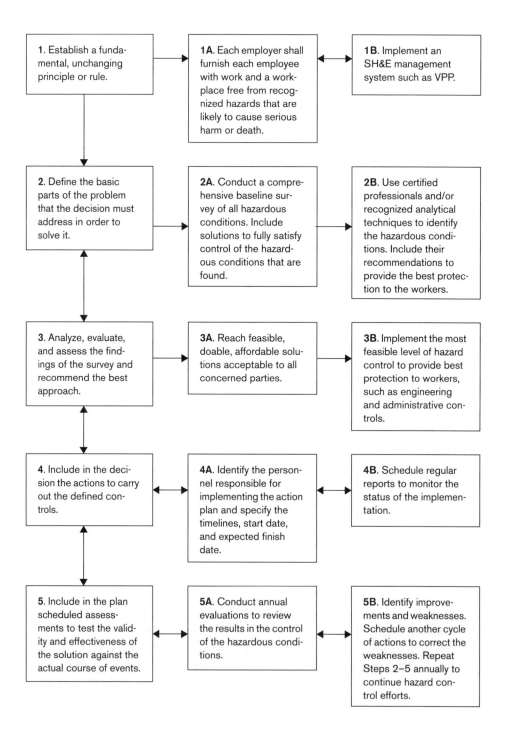

workable process that, over the long term, will be accurate, timely, and documented; and will track, follow up, and close the loop on all preventive and corrective actions. These steps apply to any problem or gap that has been found during the self-assessment.

3. **Leadership of the SH&E change must think through what solutions will fully satisfy the related conditions and parts before presenting them to the steering committee. Also, be prepared with reasons and credible justification to weather compromises, modifications, and concessions.**

In the example above, decide what is the right approach for the comprehensive baseline survey process before it is tested against all of the compromises and concessions that may be necessary to make it work in the operational areas. Members of the steering committee, especially the occupational health specialists, may have definite opinions and recommendations about the scope of the program. The financial advisers, however, may point out that the resources are not available for a program with such a broad scope; and that they may not be available for several financial cycles. Somewhere in between these two opinions, the decision must be made that provides protection adequate to meet the requirements, does not consume an excessive amount of resources, and is still close to the "right" thing to do.

This "organized disagreement,"[5] as Drucker calls it, is needed before effective decisions can be made. Leadership must be prepared to hear both sides. These discussions can also be used to defend the final decision. To defuse conflict, rebellion, or resistance, remind the dissenters that everyone's opinion was considered at the time the decision was made.

4. **Build into the decision the action to carry it out.**

Get people busy carrying out the decision at once. Get them focused on how the new SH&E management system is going to function. One of the keys to the success of changing the behavior, then the attitude, and ultimately the culture of a company, is to include in the decision the actions required to carry it out.

The most difficult mental process in decision-making was accomplished when the generic problem and its parts and solutions were specifically defined. Turning your decision into effective action will probably consume the most time because, at this stage, the mental process and the decision are transformed into reality. To facilitate the reality, the decision must have in it the specific responsibilities and work assignments for all who will be doing the work.

Be sure that the people who have to know about the decision *do* know; that they are informed of what must be done, why, who must do it and when; and that they know what the specific tasks are, so that once they *learn* the tasks they can *do* them. Requiring that a new process be performed differently than it was under the old systems gets the change in place almost painlessly. Employees eventually accept the new methods as just another way to perform their tasks.

5. **The decision should also require regular assessment that tests its validity and effectiveness against the actual course of events.**

The VPP criteria require validation and measurement of the effectiveness of the systems and functions resulting from the decisions. This is true not only in continuous activities, such as incident investigations, corrections, and follow-up; but also in annual self-evaluations when the processes and systems are measured for effectiveness and timeliness of performance against expectations. Some decisions become obsolete with time and must be changed or replaced (for instance, replacing manual typewriters with personal computers in order for the company to stay abreast of current technology and remain a viable player in the competitive marketplace).

> Executives are not paid for doing things they like to do. They are paid for getting the right things done—most of all in their task, the making of effective decisions.
>
> —Peter Drucker, *The Effective Executive*[7]

Gap Analysis Action Plan

To accomplish the tasks involved in the gap analysis, the Johnson Space Center (JSC) steering committee was divided into subcommittees according to their specialties or job functions; i.e., management leadership and commitment, hazard prevention and control and worksite analysis, employee involvement, and safety and health training. The arrangement differed from the order of the VPP; because the JSC VPP coordinators felt that it better fit the way JSC's safety and health program functioned. Also, they believed that management leadership and employee involvement were too significant individually and, if combined, would have too many tasks for one subcommittee to manage. The group was also expanded to include specialists in each of the areas, not necessarily as regular sitting members, but to be available as needed for coordination and counseling.

At the request of the steering committee, the management leadership subcommittee developed a Gap Analysis Worksheet (Exhibit 4-2) to record and track the action items. This was used when the gap was identified and until

Gap Analysis Action Plan

Exhibit 4-2. CEOSH Gap/Compliance Analysis Worksheet, JSC CEOSH Steering Committee

---

Tracking #: 9.1.2   Date: August 3, 1993
Block Leader: RH/GC
Assignee: _____

CEOSH Gap/Compliance Analysis

1. Compliance requirement: (include both references JOSHP and VPP, 1960, NHB 2710, 1910, etc.). VPP II.E.5a.(10)(C) states that "self-evaluation may be conducted by competent NASA Hqs or JSC personnel or by competent third parties…"

2. Is JSC currently fulfilling this requirement completely?
   Yes ___ No ___
   Name of Contact(s)
   What organizations(s) did you investigate?
   How did you reach this decision?

3. Are procedures documented completely? Yes ___ No ___

4. If requirement documented, attach process(s) or reference.

5a. If requirement fulfilled and not documented, describe/outline process(es).

5b. If process requires modification, describe required change or alternatives/options.

6. This is optional whether competent third parties may be used. When JSC becomes a VPP site, the OSHA site evaluation team will constitute the third party.

7. If requirement is neither fulfilled nor documented, describe/outline fulfillment options.

8. Who needs to know about this? What are your suggestions for communicating this?

[Author's note: If question 2 was answered "Yes," the item was considered closed. If the remaining questions were answered, action was transferred to the Program Development Worksheet (PDW) and this Worksheet became part of the PDW file. Over time, use of this worksheet was discontinued and all actions were recorded on the PDWs. Working Group members felt that this form was double work.]

---

recommendations were made for its closure. Then it was transferred to a Program Development Worksheet (Exhibit 4-3).

Exhibit 4-3. Sample Program Development Worksheet

Program Development Worksheet

1. Date: June 24
2. Gap Number: 2-04
3. ISHP Reference: 2.1.4.a-d
4. Responsible Person: RH/GC/CJR/CAG
5. Phone Number: 335-1607
6. ❏ New   ❏ Revision❏ Final
7. ❏ Management Commitment & Planning ❏ Hazard Assessment & Controls
   ❏ Safety & Health Training   ❏ Active Employee Participation
8. Program (Gap) Title:

    Manager Visibility

9. Description of Required Changes:

    VPP III.3.5.(5) All managers must provide visible leadership in (a) establishing clear lines of communication; (b) setting an example; (c) maintaining unrestricted employee access to top JSC management; (d) providing all workers, including contractors, equally high s/h protection.

10. Required Resources (organizations, financial, people, equipment, training, etc.):

    In place.

11. Policy and Procedures Documentation:

    In ISHP document and will be in revised JHB 1700.1.

12. Personnel/Organizational Impact:

    Impacts managers.

13. Implementation Plan (schedule and milestones):

    See On Target Schedule for Line Accountability.

14. Metrics:

    TBD

15. Follow-up Activities/Comments:

    Before this is closed, ensure that process and practices are in place for subparagraphs (a) through (d).

16. CEOSH Working Group Member Coordination: _____

    When program significantly impacts organizations external to safety and health.   *Signature/Date*

17. Gap Closure/Verification:

    Gap Number 2-04

    (cont. p. 61)

Exhibit 4-3. (cont.)

> Describe Action Taken to Close Gap:
>
> *This section should contain a complete description of the action(s) taken and rationale used to close the identified gap. Attach additional sheets if needed. In addition, you should attach copies of the pertinent procedures, letter, schedule, or whatever actually verifies that the closure has taken place.*
>
> Originator/Date   Reviewer/Date
> (Signed when submitted)   (Signed when approved)

There is no reason why the two worksheets cannot be combined. These are offered only as examples to suggest that a system should be initiated to track the gaps through the transition to closure. It is, of course, a matter of choice and convenience as to how you track the actions.

10-Step Gap Analysis Procedure

Changing from the mindset of what a safety and health program is to the mindset of what it should be takes time, persistence, and patience. Management and steering committee members must be trainers and coaches, and the committee and the employees must be trainees.

A word of caution: one of the most difficult tasks will be holding the interest of the steering committee and employees. Over time, enthusiasm may lag. The JSC VPP leadership had this problem. Other VPP participants also have it. To counteract boredom, JSC leadership recognized and celebrated small gains in a large way. VPP companies have similar celebrations. There are other methods to maintain momentum. You will read more about them in Chapter 10, "Continuing Improvement."

The JSC steering group developed a 10-step procedure from gap identification to closure.

1. **Find out what and where the gaps are.**

The majority of the gaps were identified during the self-assessment, when the current safety and health processes and systems were compared to those based on the VPP criteria.

Expect, however, that an identified gap in one area may lead to gaps in other areas that were not identified by the subcommittee responsible for those areas. This is to be expected when a company-wide plan is in the works and plans are being developed for future actions and goals.

For example, the worksite analysis subcommittee's gaps may indicate that safety and health training is deficient is some areas of the plant, but this was not discovered by the safety and health training subcommittee. The safety and health training subcommittee will then be required to find out what and where the training is needed, how hazardous is the work in those areas, how soon the training should be administered, by whom, and so on. The training subcommittee may be required to rework their schedule to accommodate these unplanned activities.

Be alert to this overlap of gaps, advise the subcommittees to expect this type of situation, and make it a rule that they work as a team to plan, communicate, and schedule the closure of the new gap. Ensure that gaps identified by each committee are reviewed by the other groups so that each can identify the deficiencies present in their own gap analysis.

2. Learn how far the gaps are from the VPP SH&E management system.

The self-assessment provides an indication of the effort required to meet the VPP criteria. The authors of *Applied Strategic Planning*[8] recommend asking five questions regarding each gap. The questions have been revised to fit the thesis of this book, but their intent remains.

- How do the desired SH&E systems and processes compare to the current ones?

  Obtain the answers by comparing the current processes to the VPP criteria as described in Chapter 3.

- How do the new goals and objectives fit with existing goals and with the current and planned resources to bring them all to reality?

  It is tempting for a planning team to exaggerate abilities and downplay what is actually possible when planning the future. Beware of this infiltrating the steering committee. Leadership is well aware of the real situation regarding bridging the gaps, how much can be done now, and how much in the future. Leadership can help the committee members contrast their wants with what is possible. Management is the anchor that keeps the strategic plan and the bridging actions from floating into the rocks of impossibility.

- What is our status toward achieving the current goals and objectives, and what does that indicate about the company's capability to meet the goals and objectives relating to the SH&E management process?

If the current goals and objectives are not progressing as planned, the projected SH&E management goals and objectives may not meet their timelines, either. Consider rethinking the current and planned goals and objectives and resources to fit reality. The answer to the question may even require that the methods for setting goals and objectives be examined.

- How does the new SH&E strategic plan (covered in Chapter 5) fit into the current strategies? Should the company's capability to accomplish all requirements be evaluated?

Take a critical look at the current strategic plan, if you have one. Is it synchronized with the company's capacities and resources, and the external realities of the marketplace? Is there flexibility in it to allow resources to be dedicated to the new SH&E management requirements? How will this new strategic plan for SH&E impact current strategies, and what modifications should be made, in which plan? Do not disregard the scope of the hazardous conditions that may take priority over all other priorities. Perhaps the planning method should be revised to accurately reflect the real life conditions and expectations throughout all of the company's operations, including SH&E.

- How will the desired culture impact the existing culture?

This is a question of critical importance. Unless leadership carefully and diligently resolves the differences between old and new cultures, the new program will not succeed. The functional structure and culture of the organization will answer this question. A values survey will indicate what changes in the existing culture must occur so that the culture and the redesigned systems fit together, as mentioned at the end of Chapter 3.

Develop the answers to these five critical questions for each gap. Whatever the answers and the decisions, direct them as much as possible toward continuing improvement and benefits for the employees and the company. Complete, or plan to complete as soon as possible, the gaps with the highest hazardous potential and the ones of significant importance. The remaining are the goals and objectives for another time.

3. **Estimate what it will take to bridge the gap in terms of money, people, effort, and time.**

Each subcommittee will analyze their gaps to estimate how long it will take to close them, if they can be closed within the time scheduled, and whether there are resources available. Each gap must have its own action plan describing the purpose, scope, and resources needed, how it can be budgeted, who is affected, who is assigned the action, in what priority order it should be completed (or if it is ongoing), and how it can be closed. If a gap can be resolved within these parameters, the subcommittee will put it into the current five-year schedule, and go on to another gap to repeat the process until all of the gaps have been evaluated against the five questions and doable or non-doable answers have been developed.

Consider other options if closing the gap does not fit within the scheduled time or requires resources that are not presently available, but is of high hazard potential or significance and must be closed now. Some options could be to extend the completion time; develop a strategy to resolve it; or decide it is non-doable at present and needs further study. In these instances, the subcommittee brings the gaps to the steering committee to modify the strategic plan for that element or to reduce the size and effort of the closure. Perhaps it can be made smaller in size or scope to make it achievable and less risky, or it can be completed incrementally over a period of two or three years.

If resources are not available, look at other programs for possible reallocation of resources to the higher priority gap, or lower the priority order of another gap to make the resources available for the first gap. Obtaining new resources is not always possible or the best option. In this era of watchful budgeting, management or the financial advisor may be reluctant to commit funds, time, or people to a new program that has no prior history of being planned, tested, or budgeted.

After all of the work that has gone into the strategic plan, it is not desirable to adjust or modify it. If all other options are not feasible, acceptable, or doable regarding one or more of the gaps, but the gap must be closed, adjusting the plan is a final option. Perhaps, the gap element, such as company culture, is too important not to be addressed. Culture will be discussed in more detail in Chapter 9, but it must be included in the gap analysis phase, since it takes a period of years to change. Or, if the gap is crucial because of its hazard potential, it must be addressed in the current year's cycle. These are too important to delay. In that case, modifications of the strategic plan are in order.

This is, after all, a continuous improvement process. If a gap cannot be completely closed in the current time frame, plan completion by next year or the year after. One of the values of self-assessment and gap analysis is to provide goals and objectives for the future and subsequent annual self-evaluations.

4. **Develop the options that will close the gap satisfactorily, but will also conserve money, people, effort, and time.**

Although costs are difficult to forecast, inflationary factors can be considered and an educated guess can be made. The time, resource, and effort questions are a part of Step 2. Here, however, consider closely the repercussions throughout the company. Management has the global view of the company's operations, budgeting cycles and practices, and market position. Management can evaluate the costs better than anyone with regard to the impact on the corporate budget, and can determine whether complete closure can be a short-term or a long-term effort.

Set time limits for regular progress reports and completion. To stay within budget and resource constraints, progress toward completion must be steady. Regular status reports will alert the steering committee to scheduled actions that are not being completed within the time allotted. Such reports can be the prompters to ask for explanations or clarifications. There can be any number of reasons for the delay or slowdown. To keep the plan on schedule, it is important to uncover the reasons and take remedial action promptly or, if feasible, reset the due date.

It could be that the action has been assigned to a committee member or an employee who does not have the authority to institute the change. It could be in a particularly sensitive area that people just do not want to confront, such as accountability. Accountability is an important element of the VPP criteria, and it is also a sensitive area not normally open to scrutiny beyond top management's offices.

This is one reason why it is within the management area of responsibility in the criteria. Management is the only entity that can initiate and approve modifications of the accountability system. The responsibility for revising accountability procedures can be delegated to a personnel resources manager, but management must not only sign off on it, they are the ones to personally implement it. When the accountability system was undergoing revision at Johnson Space Center, it was done by the Center Director's staff, but it was implemented by the Center Director at his Executive Safety Council meetings.

Ensure that the committee members working through the details of the gap closure have the authority to initiate action and follow up. If not, the subcommittee chairperson, the committee members, and the steering committee should address how best to close the gap and develop an alternative plan of action to get it done.

5. **Document the actions to provide a map of the future activities that are scheduled.**

Examine the who, what, when, and how of follow-up, verification, and validation documentation to ensure that it is specific, direct, and timely; and properly closes the "plan-do-act-check" loop. During the JSC VPP working group's gap analysis, they identified their customers or stakeholders as the different groups of personnel at JSC. The "what" were the customer needs; that is, what kind of SH&E system would fit their functions.

For example, the computer operators have different hazards and safety and health problems than the chemical laboratory technicians. The "how" involved the details of procedures, work practices, and equipment. *Applied Strategic Planning*[9] authors found that changing only one of these dimensions at a time carries the least risk of resistance to change. Changing two or three at once carries the highest risk of more intense resistance and may not succeed. For instance, with the computer operators, start with work environment first; such as an ergonomics survey to measure what equipment each operator should have to reduce cumulative disorders, then obtain the equipment and install it.

Next, or simultaneously, introduce written work practices that include training on the use of the new equipment. Finally, develop an ergonomics process to incorporate the details of why, who, what, when, where, and how. The entire set of activities could take two to three years to put in place. By that time, everyone is accustomed to the new procedure—it has become routine. The JSC group followed this method of sequential and incremental corrections. The risk of undue resistance to the change is reduced and the resources are conserved.

The length of time it takes to completely institute the new process may reduce resistance, but at the same time actions taken a year ago are sometimes not easy to recall. With a lengthy process, such as with the computer operators, a well-documented history of all of the actions will provide a paper trail for evaluations, providing resources, and determining budgetary needs.

(See Exhibit 4-2, Gap Analysis Worksheet and Exhibit 4-3, Program Development Worksheet).

6. **Consider the impact that these new processes will have on the employees who are involved in the old processes.**

The employees will be asking, "What's in this for me? What's wrong with the old way?" The communication program is the medium to answer the questions and furnish the information the employees need. Techniques must be used and messages published that convince them that this is a good move.

The communication program can be used to publicize the changes and the benefits, to explain how the employees will be involved, and to get feedback from them. As the feedback is received, the steering committee will learn of problems with the new systems and can respond. Also, feedback will identify those employees who will not or cannot adapt to the changes. This topic is discussed further in Chapter 9, "Culture and Communication."

During the gap analysis, management might discover that their leadership style is itself a gap. This is a time when management may realize that they are unwilling or unable to adjust to the challenges of these changes. There are options, of course. The decision can be made not to proceed with the VPP initiative. But leadership is convinced that this is the way the company should be positioned to meet the challenges of the future. OSHA compliance activities of the future will not be the only challenges. They will also come from the industry and national work environment.

Anyone, except leadership, who cannot adapt to the changed directions and methods, can be replaced with someone who will adapt. But when or if leadership should realize that they, personally, find it difficult to accept this new direction, someone else can be assigned temporarily to be their mentor to teach the leadership style suited to the new SH&E management concepts. The mentor can be a consultant, a representative from a VPP participant company, or a representative from the local OSHA Consultation Office. Help is available to assist in working through these challenges. As managers and leaders, they have faced similar adjustments before. After giving themselves and the VPP initiative a chance to work together, they, the employees, and the company will be rewarded by the benefits realized.

7. **After each gap has been analyzed and resolutions developed for each one, look at the entire package of closures and actions.**

Prioritize the list. Break it down to:

- Must have now;
- Need as soon as possible;
- Must have but must schedule for future budgeting;
- Nice to have but not urgent;
- Strictly a wish that may or may not materialize.

Even though the company may be on a sound financial basis, unlimited resources, time, and money are probably not readily available. The first bridges to build are those that are most critical to the state of the employees' safety and health and to the condition of the work environment.

There are also the gaps that are non-recurring, such as putting a new accountability policy and process in place. Once that it is done, it is the guide to go by indefinitely. Others, such as safety and health training, are ongoing, continuous processes without termination points. These will be particular to the organization and industry. The gaps can vary from no formal written safety and health program, no occupational health monitoring process, and no environmental conservation program, to lack of verification and follow-up documentation. Whatever the gaps are, they go on the priority list with an appropriate timeline—whether next month, next year, or three or four years forward.

With the intense effort required to close the gaps, leadership and the steering committee may ask if certain gaps can or should be closed. Those that can fall to the bottom of the list may have to await action until a later time. So long as they are not critical to the success of the overall effort, waiting their turn is not a problem with compliance. Do not fail, however, to include them in the annual self-evaluation. Keep track of them because some future change in company operations or a new OSHA rule may cause them to move up on the priority list. Or they may become outdated and can be removed from the action list.

Each subcommittee can work their priorities at the same time that the other subcommittees are working off their prioritized list. The result is that different changes will be happening simultaneously. This is continuous improvement in action.

8. **Prepare your action plans to implement the gaps.**

The steering committee and the subcommittees will develop the plans for each gap based on the analysis that they have just performed, using the guidelines in this chapter. Note that on the Program Development Worksheet (Exhibit 4-2), there is space for actions taken to close the gap.

9. **Once the steering committee and subcommittees have completed action plans, implementation begins.**

This is covered in Chapter 6, "Implementation of the Model SH&E Management System."

10. **Track the gaps until they are completed, are in place, and are performing as expected.**

The steering committee continues to receive regular progress reports until the gaps are completely closed.

The tracking may be over a period of years with some gaps. Make these reports part of the regular reporting schedule.

More on this topic is in Chapter 10, "Continuing Improvement." The ongoing closure activities become goals and objectives for the future.

Summary

In this chapter you have been led through the steps of gap analysis to structure and organize a plan of action that will affect company-wide operations. In this case, the action is adapting the safety and health management systems to a new set of criteria. Pointers on effective decision-making are also offered to assist the steering committee in making the complex and, sometimes, painful choices that accompany a company-wide internal change. Ten steps of gap analysis are discussed to provide assistance in developing a plan of action to close the gaps. Examples are included of the worksheets used by the JSC group to track their action items. Chapter 5 takes you through a strategic plan for closing the gaps.

References

1. Goodstein, Leonard, Timothy Nolan, and William J. Pfeiffer. *Applied Strategic Planning: a comprehensive guide.* New York: McGraw-Hill, 1993: 261.
2. Drucker, Peter F. *The Effective Executive.* New York: Harper & Row, 1967: 192.
3. Ibid: 104.
4. Ibid: 123.
5. Ibid: 153.
6. Based on Drucker, Peter F. *The Effective Executive.* 123–127.
7. Drucker, Peter F. *The Effective Executive.* 158.
8. Goodstein, Leonard, Timothy Nolan, and William J. Pfeiffer. *Applied Strategic Planning: a comprehensive guide.* New York: McGraw-Hill, 1993: 264.
9. Ibid: 271.

Where Do You Want to Be?

5
How to Get There From Here
The Strategic Plan

> ...we define strategic planning as the 'process by which the guiding members of an organization envision its future and develop the necessary procedures and operations to achieve that future.' This vision of the future state of the organization provides both a direction in which the organization should move and the energy to begin that move.... It involves a belief that aspects of the future can be influenced and changed by what we do now...the strategic planning process does more than plan for the future; it helps an organization to create its future...
> —Goodstein, Nolan, and Pfeiffer[1]

Your SH&E program now has a new focus and a defined goal. The focus is to bring the company's SH&E management system up to the standards of the VPP criteria. The goal is zero injuries and illnesses. However, the task is complex. Each operation, function, and department has different tasks, work environments, supplies, and equipment. Much of it is the same—for instance, everyone uses pens, pencils, clips, paper, telephones—but much of it differs. Offices do not work with the same equipment as the machine shops or the warehouses. Each function has its own hazards, risks, precautions, and controls.

Now the task is to bring diverse programs and complex operations under one umbrella SH&E system. All top managers and company leaders have a vision of the company's future. Within their assigned areas of work, they have the responsibility to realize that vision. This is true of almost all organizations that plot new courses. There are general similarities; but each will have individual differences.

Why a Strategic Plan?

As the quotation at the beginning of the chapter suggests, strategic planning not only predicts your future, it also helps to create it. You and everyone around you are gifted with creative imaginations. That's how we plan future meetings, how we set goals for ourselves, and how we draw plans for our new home—we are creating our future reality. You can do the same by planning the future state of the SH&E management system.

Strategic planning integrates all of the business functions and cycles. It is for the future; yet maintains the status quo of the present. Strategic planning provides the company with a map for today and tomorrow. The map now requires rerouting to include the tasks necessary to achieve improved processes. The strategic plan is the route to your destination. The challenge is to get there from here.

To successfully follow a new direction and reach the desired destination, the plan must be specific and clear, always keeping track of where it is now and where it is to be in the future. You know the route, the cost, the means of travel, the time frame, and the accommodations when you reach the destination. Now integrate all of these factors into an organized, cohesive action plan.

The number and nature of the complexities will vary from one organization to another, but be assured that challenges will be present to some degree in every organization. To help you get prepared, here are a few challenges to consider.

Challenge 1: Resistance to company-wide change

Any change that affects all of the company operations causes the ground to shift under everyone. Resistance to the change is a common reaction. The current status may not be the most desirable; but change to a different status awakens a fear of the unknown in the people affected. The changemakers must counteract these fears and anxieties by overt and positive actions. Those actions must be defined and planned for the various stakeholders.

Challenge 2: Safety and health issues affect and involve all operations and departments

The criteria get into all operations and departments. This may be new for some of the managers and department heads. The present attitude may be, "Safety is safety's business; not mine." Leaving safety and health issues to the SH&E staff is the normal routine for many companies. Indifference and neglect of the SH&E process exist in many organizations. Lip service at the appropriate times is practiced regularly at the top and middle management levels. Stonewalling among managers and department heads frequently occurs where change of any nature is involved. The new focus pries them out of their comfort zones and threatens their mini-empires. The reactions may extend from cynical indifference to covert sabotage. One of leadership's first tasks is to recognize that these reactions may occur, to have a plan to overcome them, and then to gain the support and commitment that is essential to the system's success.

Challenge 3: It takes time to get there

The gap analyses have finished the estimates of how long it will take to reach VPP quality. The long-term activities will take a few months or a few years. Most companies that participate in the VPP model program take from three to five years getting ready to qualify. The first two years are typically spent changing the paradigms of the managers and the employees so they accept, support, and get involved in the new program.

If this is a voluntary effort and the leadership does not intend to apply for VPP participation, you have the advantage of scheduling implementation of the criteria to fit into the overall company planning. Another advantage of the VPP requirements is that the company is required only to *start planning* the inclusion of the critical elements and to *correct the most hazardous conditions at once*. The remaining requirements can be implemented incrementally to fit a company's time and budget schedule. Keep in mind that continual improvement is one of the primary expectations of the VPP criteria, so step-by-step corrections over a period of time are acceptable, just so long as the most hazardous, hurtful gaps are first priority.

In 1992, the Johnson Space Center (JSC) steering committee plotted a four-year plan (see Exhibit 5-1). This is a normal timeline. The application process was originally scheduled for 1994. Due to external circumstances, the plan was set back to 1998. Contingency planning should be a part of the plans to manage unexpected setbacks and obstacles; but a projected timeline of four to five years is realistic.

Challenge 4: The program must have involvement and commitment from all employees

For this new program to succeed, everyone who gets a paycheck must be involved. This means that all employees—top managers, middle managers, senior lead people, and other employees—accept their *personal* responsibility and accountability for the safety and health program. A strategy is needed to inform and educate all levels of employees in the company and create an awareness of what the SH&E system is about.

Awareness is only the first step toward accepting responsibility and accountability. To do this, there are several options. Communication specialists are available to mount an active publicity campaign to sell VPP to the employees. Based on the size and complexity of company operations, an outside consultant may assist in developing the communications effort, or the in-house public relations department or person can do the job.

Exhibit 5-1. Phase One Project Management Plan

Task Name	Duration	Start	Finish	Resource Name	1992 1 2 3 4	1993 1 2 3 4	1994 1 2 3 4	1995 1 2 3 4	1996 1 2 3 4
1.0 STRATEGIC PLAN	276 days	08/11/92	08/31/93	SM/TM		SM/TIM			
2.0 GAP ANALYSIS	158 days	01/04/93	08/11/93	Sgp Leaders		Sgp Leaders			
3.0 PROG PROJ DEVELOP	262 days	08/11/93	08/11/94	RH/GC			RH/GC		
4.0 PROGRAM IMPLEMENT	262 days	08/11/93	08/11/94	RH/GC			RH/GC		
5.0 APPLICATION PROCE	66 days	08/11/94	11/10/94	RH/GC				RH/GC	
6.0 EVALUATION	67 days	11/10/94	02/10/95	RH/GC				RH/GC	
7.0 IMPLEM DOL FINDINGS	67 days	11/10/94	02/10/95	RH/GC				RH/GC	
8.0 EVALUATE & ADJUST	262 days	02/10/95	02/12/96	RH/GC					RH/GC

Some means of ensuring all-hands awareness and involvement is an essential feature to include in the strategic plan.

Challenge 5: The program may require long-term budgeting for compliance

The preceding four considerations are concerned mostly with people matters—getting the employees to accept and participate in the program. Long-term budgeting addresses the material matters—new equipment, modified work areas, and modifying some processes because they do not meet the SH&E criteria. Once the baseline self-assessment and the gap analysis are completed, do not be surprised if you discover that some of the current processes are just not acceptable under this new set of rules. To modify them into acceptable and compliant systems may take a considerable outlay of capital. Whether the money must be spent or other less expensive options are available that still meet the requirements will depend on flexibility, creativity, and management's overall plans to grow the company. What is done or planned for the SH&E program should be included in the long-range plans for the company.

Keep in mind that OSHA is open to creative, innovative methods to meet the criteria. The criteria do not state that the company must accomplish all of the redesigning and modifications in any set period of time. What OSHA does require is that the company recognize that redesigns and modifications are needed and have been identified, and that a plan is developed to get the related tasks completed in a predetermined period of time. OSHA looks at the priority that has been assigned and evaluates the timelines against improvement in the environment to enhance the safety and health of your employees. Since the company has committed to the program and intends to meet the criteria voluntarily, it is paramount that the requirements are met with solid, comprehensive efforts. When the next annual self-evaluation compares the company systems and processes to the criteria, leadership should anticipate a straight-A report. This is the continuous improvement and motivational aspect of the criteria.

Challenge 6: Provide a unifying focus for all stakeholders

All of the actions taken to get the strategic plan developed, prepared, and put into action can serve as the medium to get the planning group working like a team. The strategic planning actions can include all of the managers and employees with the new direction for the SH&E program and overcome some of the obstacles mentioned in the preceding paragraphs. Get a group of people focused on a new challenge, and a win-win synergism develops. Leadership's personal challenge is to make the right moves to get people focused and enthusiastic.

The JSC steering group used a classic model of the parts of a strategic plan, the future conditions and circumstances to anticipate, and Goodstein, Nolan, and Pfeiffer's how-to's for developing a strategic business plan. (See their book, *Applied Strategic Planning: How to Develop a Plan That Really Works, A Comprehensive Guide,* for complete details.) The JSC steering group spent the first year developing the agenda and strategic plan for the VPP initiative. What follows is an overview of the activities that engaged their attention.

The Pre-Plan Agenda to Activate the Strategic Plan

A comprehensive agenda is needed to organize the details of the who, what, when, where, why, and how of the plan. A trustworthy, knowledgeable steering group is a necessity to write the agenda, to prepare the plan, and to guide the company through the transition into the final steps of implementation.

Form a steering group of key stakeholders

The steering group is already in place. These are the people who were selected to discuss the first major decision: whether to use the VPP criteria to go for excellence in the company's safety and health program. They were also the ones who conducted the self-assessment.

Several weeks before the full JSC steering group was formed, the JSC-designated SH&E management met with some of the key players to discuss the feasibility and acceptability of raising the level of the Center's safety and health program to the VPP requirements. The first action was to have one-on-one meetings with key JSC people, such as the head of the occupational health program, whose support was essential for the health side of VPP. Meetings were held with representatives of the largest contractors on the site, such as Rockwell, Lockheed, and Boeing, and the government employees' union representative. First on the list was the Director of Safety, Reliability, and Quality Assurance, who had to be convinced that there were real benefits to accrue to JSC if the VPP initiative was achieved. The Director immediately saw the external and internal benefits. His comment was, "It's going to be a lot of work, but go for it." He was a key stakeholder, but not an active member of the steering group. The designated Director's representative kept him updated with regular status reports.

Your company may not have such a variety of players or a hierarchy of people. It is always a good idea, however, to get key players on the team and enlist their support *before* broaching a new program that will involve everyone. The final, enlarged JSC steering group consisted of representatives from the safety and health offices and their support contractors. Rockwell, Lockheed, and a facilitator/strategic planner completed the group. The facilitator started the strategic planning and led the first meetings to get the group pointed in the right direction.

After the group got into the safety and health program planning and evaluation phases, the outside facilitator was no longer needed. During this later phase, the group was enlarged to include other contractor and civil service personnel and to form sub-groups for each major element of the VPP criteria.

From time to time, other experts were brought in for special topics, such as human resources, labor relations, and public relations. It is also worthwhile to have an outside expert advisor and mentor who can advise and counsel you concerning the fine details of implementing the criteria. Mentors usually are people who work for a VPP participant. When a company is approved as a VPP participant, OSHA encourages the company to make a representative available for coaching similar companies in the implementation and approval processes. Mentors can be located by contacting the local OSHA office or the Voluntary Protection Programs Participants Association (www.vpppa.org/index.cfm). There are also consultants available who are experienced in the process. The VPPPA mentors usually do not charge for their efforts. With the mentor added, this is the core group that was the beginning of involving all the JSC employees.

Your company's steering group might consist of representatives from production, marketing, purchasing, financial planning, human resources, public relations, unions, safety and health, or any other combinations of essential departments. It can be as large or small as wanted or needed, depending upon company size and complexity. It is also important to select people who are liked and respected and who work well in groups. Avoid, if possible, employees who have large egos and tend to dominate meetings or who are the strong resistors to the new SH&E management approach. They may work from within to undermine and sabotage the new initiative. Select people who work unselfishly for the general benefit of the worker and company.

The corporate designee and chairperson of the management subcommittee can facilitate these meetings or have one of the group act as facilitator. Whether the company engages an outside consultant to help in this process depends on the company's inclinations and business practices. The mission of the group will be to develop a strategic plan to get the company's safety and health program to VPP Star quality within a stated period using the OSHA VPP requirements as the criteria.

Consider steps to maintain corporate infrastructure through the transition from the old to the new

The management subcommittee knows better than any others in the company how the organization is structured, what the long-term plans are, in what direction the leadership is taking the company, and how it intends to remain a viable competitor in the market. How the company moves toward implementing the new program will depend upon its culture.

If it is a formalized culture that is built upon conservative risk-taking, the approach to the VPP initiative will be quite different than if it is flexible to change with changing times. With the former, movement forward will be slow and cautious. With the open, flexible work environment, employees will be better conditioned to try this new approach to SH&E without too many disturbances within the infrastructure. Departments may or may not remain the same; old ones may be reorganized; new ones may be added. Staff may shift from one function to another, and the occupational health program or the safety program may require additional staff.

The identification of these changes will occur during the gap analysis, but for strategic planning purposes, they can be phased into the company structure at the opportune time to allow for convenient budgeting and easier acceptance. How this takes place is the company leadership's decision. During the gap analysis and the strategic planning process, however, is the time to identify where and what the changes are, how they will be accomplished, when they will take place, and what the projected cost is. Get the dominoes standing up in the order in which you want them to fall.

Safety and Health Mission Statement

As one of their first actions, the JSC steering group wrote a mission statement. What does a mission statement have to do with this management system? At the start of a new endeavor that is going to be comprehensive and will have a long-term affect on the entire organization, it is a good idea to focus at the beginning on the why, what, where, when, and how of the endeavor. Answering these questions specifically gets the group focused on their purpose—what they are trying to accomplish, why they are trying to accomplish this particular effort, how they plan to do it, and what the benefits will be.

The act of writing a mission statement helps to diffuse objections and differences and serves as a middle ground on which to reach agreement. It is also a medium to encourage the group to work together as a team. Be sure, however, that your company's mission statement for the safety and health program supports the company's mission. The two should be in alignment.

The JSC steering committee wrote this mission statement:

> Our mission is to enhance the safety and health of all employees by establishing the Johnson Space Center as a Center of Excellence for Occupational Safety and Health (CEOSH), and by maintaining a process of continuous improvement using the external criteria of OSHA's Voluntary Protection Programs (VPP) as a benchmark.

As published July 2002 on their website, ExxonMobil's commitment to safety, health, and the environment was:

> ExxonMobil is committed to maintaining high standards of safety, health, and environmental care. We comply with all applicable environmental laws and regulations, and apply reasonable standards where laws and regulations do not exist. Energy and chemicals are essential to economic growth, and their production and consumption need not conflict with protecting health and safety or safeguarding the environment. Our goal is to drive injuries, illnesses, operational incidents, and releases as close to zero as possible.

ExxonMobil recognizes the need to align the SH&E goals and objectives with the need to recognize and control, as far as possible, the hazards that their products present to the environment. (As an aside, at that time according to the VPPPA website, ExxonMobil had 13 Star VPP sites.)

Examine the Corporate Culture

Examine the corporate culture and beliefs of all company members and external stakeholders. Identify differences and foster consensus for the new SH&E management system. Success or failure depends upon how the employees receive the message and what their responses are to the ideas of empowerment and participation. This is a critical step before strategies are developed to implement the new SH&E management system.

In October–November 1992, JSC tested the water by conducting an informal survey among the civil service departments (called directorates in government organizations) senior staff, branch chiefs, and safety representatives, and the contractor managers and safety representatives. This group represented a cross-section at the Center. Exhibit 5-2 is a summary of the responses regarding the strengths and weaknesses of the JSC safety and health program.

Exhibit 5-2. Summary of JSC Safety and Health Program Survey

Top 10 Strengths	Top 10 Weaknesses
1. Access to medical services.	1. Emergency training drills.
2. Accident investigation and reporting.	2. New-hire training.
3. Access to safety and health professionals.	3. Supervisors' safety training.
4. Top management commitment.	4. Job and process hazard analysis.
5. Line management commitment.	5. Written safety plans and procedures.
6. Employees can express concern without reprisals.	6. Line management's responsibilities for health and safety are documented and communicated.
7. Contractor personnel have the same level of protection as civil servants.	7. Obtaining necessary resources.
8. System for identifying and correcting hazards.	8. Methods to ensure adherence to safety rules.
9. Employees can air concerns to management.	9. Safety planning integrated into overall management system.
10. Emergency response plans.	10. Employee involvement.

The survey contained 31 statements that respondents answered on a scale of one to five. A response of one indicated "Strongly Disagree" and a five represented "Strongly Agree." All statements were worded so that a "Strongly Agree" response was a positive. The statements were divided into five main areas:

- Top management commitment (6 statements);
- Line management commitment (6 statements);
- Employee commitment/Participation (4 statements);
- Hazard assessment (8 statements);
- Safety procedures and training (7 statements).

The statements were taken from the OSHA VPP checklist of requirements for participating organizations.

Of the 220 survey questionnaires distributed, 119 were returned. The returned surveys indicated that about half of each group responded. The groups were division (or department) heads, civil service safety representatives, contractor program managers, and contractor safety representatives.

Random sampling of comments by respondents

Contractor manager:

- I am not personally familiar with some of the questions.
- Overall, I believe JSC and the contractors do very well in safety and health areas.

NASA division chiefs:

- The JSC safety and health program far exceeds the programs in the commercial community.
- ...generally speaking, there is very little attention or benefit from JSC programs.

NASA safety representatives:

- Other than safety reps, there is little or no training for safety at NASA for new hires or other employees. JSC needs basic safety classes for all employees and maybe a refresher class every year...
- Most of the questions which I answered with "3" were questions for which I wasn't sure of the answer.

Contractor safety representatives:

- Necessary resources are not provided. For example, building XX needs an external chemical building. Contractors do a much better job than civil service, especially in line management commitment, employee training, and new-hire training.
- Too much of the safety mentality is left to the safety professional. Supervision has not fully accepted a leadership role in the process.
- This survey was difficult to answer due to the fact that I have limited insight to JSC management and civil servants' exposure to the safety programs. I also feel there is a large gap between line supervisors and top management communication because it seems apparent that employees on the floor are not always knowledgeable on safety procedures.

One of the statistical analysts at the Center evaluated the answers. Here are some of the findings:

- NASA and contractor safety representatives rated the JSC safety and health program lower on the average than the department heads and program managers.
- Top management commitment had the highest overall average.
- Safety procedures and training had the lowest.
- Line management commitment and safety procedures and training had the largest number of unfavorable responses.
- The other three sets were close to each other.

The JSC VPP management coordinator found what he suspected:

- That top management commitment was evident; but line management commitment was not.
- That the safety representatives who were closer to the day-to-day working of the current safety and health program were more critical and more perceptive of its strengths and weaknesses.
- That training was weak in several areas.

This survey gave him some foundation for planning an approach to communicating the idea to the various levels of management and employees. How it was done is covered in Chapter 9, "The Stakeholders: Culture and Communication."

Whether you use JSC's approach or another to inquire into the beliefs and culture of the employees, it is necessary to discover by some means the true state of the work environment and feelings of the workforce about SH&E. One of the VPP prime features is the involvement of the total employee population. In informal surveys and in the formal evaluations conducted by OSHA, they have found that employee involvement is one of the significant reasons for VPP success.

Address obstacles as they occur during the change

During the self-assessment, gap analysis, strategic planning, and implementation periods of transforming from the old to the new SH&E management system, unexpected situations and conditions will occur. For example, the union may be expected to support the proposed new system because it totally involves the employees in deciding what their safety and health activities should be. But instead, the union may resist the idea. They may think company leadership is making a power play to usurp their control, or it may appear to be a threat to their ability to use safety and health conditions as a bargaining chip, or there may be a lack of trust or bad feelings between union and management somewhere in the organization. This situation calls for a contingency plan to overcome the union's resistance.

Expect some surprises similar to this one and be prepared to handle them as they arise. Use these generic steps to analyze the unexpected scenarios:

- What steps must be taken to reach the solution;
- Who will be responsible;
- An estimate of the time frame to completion;
- How the company culture must change to support the new conditions.

Evaluate the scenarios during the self-assessment and the gap analysis. Then you will be prepared to take appropriate actions to achieve the criteria expectations. During the self-assessment and the gap analysis, you will also find out how far you are from the destination and how many bridges have to be built to get there. Depending on the extent, timelines and strategies can be projected to reach the destinations. Exhibit 5-3 is an example timeline chart.

Perform a company-wide self-assessment of current safety and health systems using OSHA VPP criteria as benchmarks

To conduct this first assessment and ensure that all of the requirements that OSHA expects have been considered, use the pre-approval evaluation method of the OSHA VPP team (Appendix D, "Guidelines Checklist Used by VPP Site Evaluators") when they visit a plant that has applied for acceptance into the VPP. It is thorough and comprehensive; and it will produce a true reading of where the company program is in comparison to the criteria.

Chapter 7, which discusses self-evaluation of the new SH&E system in detail, is based on the OSHA VPP evaluation procedure. The first self-assessment of your current safety and health program activities should be as thorough as the VPP team would be, so there will be glaring omissions of essential parts when you begin implementation.

A note is in order, however, about the self-assessment covered in Chapter 4. The preliminary discovery in Chapter 4 does not include the complete three-step method that OSHA follows for site evaluations. The self-assessment in Chapter 4 provides enough information for you to gauge how far or how close the company's SH&E program is to meeting VPP criteria. Chapter 7, however, describes the entire evaluation process used

Exhibit 5-3. Critical Element Schedule

ID	Task Name	Duration	Start	Finish	Resource Name
1	Written S/H Program	388 days	1/6/93	7/1/94	CAG
2	Review 1960/VPP	6 days	1/6/93	1/13/93	Group
3	Assign chapters	1 day	1/13/93	1/13/93	Group
4	Table of contents	19 days	1/13/93	2/8/93	Group
5	First draft	11 days	2/8/93	2/22/93	Group
6	Group review	1 day	2/22/93	2/22/93	Group
7	Incorporate comments	26 days	2/22/93	3/29/93	Group
8	Ron review	48 days	3/29/93	6/2/93	RAM
9	Incorporate comments	7 days	6/2/93	6/10/93	Group
10	ND4/SD26/SR review	23 days	7/15/93	8/16/93	Branch/SRs
11	Incorporate comments	28 wks	8/4/93	2/15/94	CAG
12	Final copy	40.2 wks	8/25/93	6/1/94	RAM
13	Policy/goal/objs	30.4 wks	1/7/93	8/6/93	RHH/GCC
14	JSC plan include S/H	365 days	1/7/93	6/1/94	RHH/GCC
15	Responsibility/authority	73 wks	1/7/93	6/1/94	RHH/GCC
16	Visible leadership	45.4 wks	8/19/93	7/1/94	RHH/GCC
17	Include in safety manual	45.4 wks	8/19/93	7/1/94	CAG/RAM
18	ISHP technical memo	45.4 wks	8/19/93	7/1/94	CAG/RAM
19	Publish	43.4 wks	9/2/93	7/1/94	Mgmt Svcs

by OSHA for approval and continued certification purposes. It is recommended that the VPP evaluation procedure be used for annual evaluations.

Identify the gaps between your safety and health program performance and OSHA's program criteria

This was addressed in Chapter 4. During this period, there will be some actions that can be completed quickly and need to be done only once. Other actions may take months or years. The strategic plan should include tracking and progress reports until actions are completed. The JSC steering committee divided these tasks for each critical element according to their sub-groups and tracked them by the timeline program (see Exhibit 5-3).

The gap identification should be accompanied with a detailed plan for procedures, budgets, human resources, and timelines. At this point, more people are frequently included to cover these details. The steering group members can be the contact people, but the employees in each of these areas will be needed to provide input for each item. Paralleling the institution of the new work environment, parts of the old system will continue to operate until the new system is fully functional and efficient. Each area of the VPP criteria requires its own set of plans, procedures, budgets, human resources, and timelines. All of the elements of the system are functioning in conjunction with each other, but not all elements are at the same stage of development at the same time. In addition, the new elements will probably not go into operation at the same time. For example, you may not have an adequate new employee orientation program. Neither is there a senior staff safety and health training process. These two plans are part of the training element of the criteria; but the new employee orientation program should be integrated before the senior staff training. The worksite analysis and hazard prevention and control elements each have several requirements. Phase in the critical requirements of each first and schedule the others in sequential order.

Develop a company-specific contingency plan to recognize internal or external factors that could dictate changes in the overall plan

While the steering committee is engaged in assessing and strategizing for this new SH&E management system, other company business matters also require attention—production, marketing, distribution, human resources, budgeting, and others. Management is also watching the external factors, such as economic conditions for the industry and the country; activities of competitors; and federal, state, and local governmental affairs, especially what OSHA and EPA may be doing, that will affect the company and the SH&E system.

As the strategic planners do, it is recommended that a contingency plan be included in the strategic plan to accommodate changes in these external forces that could result in adjusting the course. Even a change in the political arena, whether Republican or Democrat, can change the government's attitude toward compliance, and result in adjusting some area of the SH&E management system. As pointed out earlier, the codification of the VPP criteria is a topic in Congress that is awaiting its turn to be heard after more immediate issues are addressed. No one knows what the outcome will be and whether it will be beneficial or negative for businesses. Be aware that such events may occur and be ready to respond proactively to optimize whatever advantage is available.

Integrate the strategic plan along with the VPP critical element requirements into the operational areas of your company.

At this stage, you have identified and planned for the most significant factors and are ready to implement the SH&E management system, which is covered in Chapter 6. From here it is a matter of tracking the changes and evaluating the entire system annually to attain and maintain the excellence in SH&E performance.

Summary

Chapter 5 discusses the why's and how's of a strategic plan, how to start, what it should contain, and what the results should be. The strategic planning of the VPP steering group of the Johnson Space Center, NASA, was described as a case study of some of the techniques that can be applied. Examples are provided of the scheduling tables that were used by the JSC group. The chapter emphasizes that the company is transitioning from one set of criteria to a new set with much detail and several challenging requirements. Leadership is encouraged to include contingency planning in the strategic plan to be prepared for unexpected internal or external events that can affect the company status and sidetrack the strategic plan.

Chapter 6 covers implementation activities of the updated SH&E management system. Chapter 7 walks you through the first annual self-evaluation that will be conducted using all of the VPP evaluation methods. Each one is an important step on the high road to excellence in the SH&E management system.

References

1. Goodstein, Leonard, Timothy Nolan, and William J. Pfeiffer. *Applied Strategic Planning: a comprehensive guide.* New York: McGraw-Hill, 1993: 3.

6 Implement the Model SH&E Management System

...a military organization does not have to be very large before those who fight must get their commands from somebody who is far from the scene of combat. This means that there has to be a 'plan,' a preparation for carrying it out, and preparation for changing it if necessary. If the plan is changed fast or without preparation, total confusion ensues. Some of the people at the front will still follow the old plan and simply get into the way of those who have switched to whatever the new plan demands. The bigger the organization, the longer it takes to change direction, the more important it is to maintain a course...

—Peter F. Drucker[1]

The stage has been reached in the strategic planning where you are ready to develop action plans for integrating the SH&E processes into the company's functional and operational units. Action plans were discussed in Chapter 4, "Where Do You Want to Be? Gap Analysis," giving you the when, what, and how.

Integrating Action Plans

The first step is to ensure that all of the action plans fit into the corporate plans and with each other. The strategic planners Goodstein, Nolan, and Pfeiffer call this "integrating action plans—horizontally and vertically."[2]

So far in this process, you have different action plans that identify the gaps that must be bridged or closed company-wide according to the critical elements of the criteria:

- Leadership commitment and employee involvement;
- Worksite analysis;
- Hazard prevention and control;
- Safety and health training.

These cover the spectrum of company operations and are not segmented according to duties, tasks, and environment in each work area or department.

The vertical action plans for each operation or unit identify exactly what training, costs, equipment, structural modification, and other resources are required; what is the immediacy of bridging the gaps; who outside of the operations and units will be affected; and any unique gaps for each area. For example, you need to know what safety and health training specialists are needed for all of the operations and units, such as welding, fall protection, forklift and crane operations; or ergonomics, monitoring, and industrial hygiene.

Integration of these action plans involves identifying the commonalties of the elements and the priorities horizontally. For example, much of the basic safety and health training can be done horizontally across the company. Hazard recognition should be identified as a safety and health training requirement for all employees, but the training can be separated according to the level and nature of hazardous work that each department or operation performs.

Workers in the heavy industrial or process unit environments, such as machine operators, pipe fitters, maintenance crews, welders, warehouse staff, vehicle operators, and crane and forklift operators, perform a higher level of hazardous work than computer operators and office support staff. So even though all will receive the appropriate hazard recognition training for their particular level of hazardous work, the higher hazard exposure training should be scheduled first. The same rationale applies to other training, such as worksite analysis, and hazard prevention and control. Computer operators and office staff still must know how to perform worksite analysis, and what prevention and controls should be applied in their work areas, but training is not as immediate to their safety as it is for the industrial workers.

The vertical integration involves the same rationale, only according to levels of employees. Refer to the communication circles displayed in Exhibits 9-1 and 9-2. They are divided according to different departments and to functions. The safety staff, if you have one, is usually the group that coordinates training needs. Sometimes they conduct the training; sometimes specially-trained teachers are hired temporarily. The vertical integration occurs functionally. The supervisors of the warehouse not only need the same training and experience as the machine shop workers, but also additional training in their SH&E management tasks.

Also note in the circle that the JSC plan was to educate and train the safety staff first, then the senior staff, then the managers and division chiefs, and then the grassroots. It was not done exactly in that order. Sometimes training was being given to all these groups, but it was different for each group. The grassroots employees might have been attending a safety and health fair at the JSC recreational center, while at the same time the senior staff were attending a DuPont safety course for senior managers and supervisors.

The senior staff, managers, and supervisors will receive essentially the same type of training except for the level of responsibility. The senior staff across the functional lines will receive executive overviews of their duties and responsibilities. Middle managers will receive more detailed training regarding tasks, duties, and responsibilities. Supervisors of grassroots employees will be trained in the duties even more specific to their jobs; and grassroots training will concern work practices in their area, vertical just for their departments. Some training is horizontal since many work practices are used across the board—housekeeping, slips and falls, lifting, and office. Some are specialized—laboratory, research areas, and warehouse. You should conserve resources whenever possible and provide the same training horizontally to several functions.

What is important is that the action plans are integrated and coordinated, vertically and horizontally, so that when the plans are completed, they will define precisely who, what, when, where, why, and how the SH&E processes and systems will be implemented as a whole grand plan, working and fitting together. The action plans should include at least:

1. A precise description of what the gap or requirement is.
2. Where it exists and who it is for.
3. Why it is necessary.
4. The people, facilities, machinery, and equipment necessary to build the bridge and the costs.
5. A milestone chart for completion from start to finish.

Some of the plans will be complex and some will be simple. Some will take only a one-time action; others will take years with several phases. The Project Management Plan (Exhibit 5-1) is an example of how the plans can be presented briefly and concisely.

Be certain everyone is on board

Someone has to be the engineer to ensure that all of the cars in the train are hooked together properly and in order, and that everyone is on board taking care of his or her tasks. In this initiative, company leadership is the engineer. Strive for consensus among the vertical and horizontal functional areas. One function must agree to and with another's action plan.

Beware that they do not start competing against each other for the resources that may be available. The VPP criteria foundation is values-based. Allocation of resources must be in alignment with the critical, life-saving gaps versus others, such as databases and statistics gathering. SH&E leadership will have to agonize over the priorities, timelines, and costs to balance the tasks among the variety of jobs that must be done in the SH&E arenas, as well as in the other company operations and functions.

Watch that certain work areas are not overloaded. There may be tasks and duties that must be performed for the new processes that require the attention of staff or operational members who are already burdened with their usual duties. For example, the comprehensive safety and health baseline survey to identify the existing or potential hazards, wherever they may be, is an extensive task. To reduce the magnitude of the effort, the survey can be divided according to areas—for instance, laboratory, office, or machine shop—but the support and technical staff in some of these areas may already be loaded with work. In these instances, use task forces composed of support and technical employees who can devote a specified time to plan, do, and prepare the survey reports in their areas. One or two members on your steering committee (again, depending upon the size of your company) can be the coordinators to collect the reports and consolidate them into one. That job is then done until the next year when it must only be updated.

Setting priorities

Be certain that priorities and timelines are included in the action plans. Time is on your side—normally strategic plans are in five-year cycles. This time frame allows leeway to stretch out the priorities for completion. Five years also provides time to budget for the bridges that must be built. Be sure that senior staff and managers know that not all of their programs and processes will be done at once. Make them aware that they must use a realistic yardstick to accomplish their objectives. This especially applies to small to medium-sized companies, since resources may not be available for all at once.

When you set priorities to close the gaps, you are confronted not only with the internal scheduling and resource problems described above, but also with the external pressures exerted by OSHA's deadlines to be in compliance. (OSHA sometimes sets priorities for you.)

When OSHA issues new rules or revisions to existing ones, they usually carry an effective date by which the employer must be in compliance. Depending upon the size of the company and the extent of resources, OSHA will normally allow reasonable time to come into compliance. Unless it is a life-threatening condition in the plant or company, OSHA can accept as reasonable an abatement plan that describes what will be done by certain dates, not necessarily the date stated in the rule. This, of course, will depend upon whether the hazards being controlled have a high or low potential for severity. If you show good-faith intention to be in full compliance by including abatement plans and budgets in the five-year strategic plan, OSHA will know that you are aware of the rule, its requirements, and its compliance deadlines. Since

OSHA can knock on your door any day without notice, it is prudent to either comply with the rules by the required dates or be prepared with an abatement plan that OSHA will find acceptable.

Keeping in mind the severity of the hazard potential requiring compliance, the least severe can be lower on the priority list than the most severe, and can usually extend further into the future. If compliance involves the fall protection rule, consider this as controlling a life-threatening and imminent hazard and take immediate action, even if another priority has to drop down on the list. If compliance involves the process safety management rule, put it high on the list but lengthen the time required to come into full compliance. There are trade-offs and options in the compliance process to allow for realistic priorities.

The significant point is that you do have a plan with your goals, objectives, and budgets articulated to improve the SH&E work conditions, and that the strategic plan for SH&E management systems is integrated into the overriding strategic plan for the company. This integration into company-wide planning actively meets the VPP criteria that requires the inclusion of SH&E considerations into corporate strategic planning cycles, along with the marketing, human resources, production, services, budget, and other processes.

It is important, also, that SH&E subcommittees take into account what new people may be needed within the next two to four years for the new processes and systems as they are implemented and mature. Do not get trapped in a situation where the process is successfully functioning, and a point is reached when someone is needed to attend to it full time and that someone is not available.

Other processes will also experience these growth developments. Give thought to combining the coordination duties for more than one process into one or two jobs, and plan to have the financial resources available to meet this need. Depending upon the company size, it may be large enough that one person per process is required, or small enough that one person can coordinate all processes. These are individual company choices based on what resources are available. Ensure that leadership and the steering committee are ready to begin implementation before it actually begins—which brings us to contingency planning.

Contingency Planning

Contingency planning recognizes that an unexpected threat or opportunity may occur any day. The authors of *Applied Strategic Planning* describe it this way: "Contingency plans are preparations for specific actions that

Implement the Model SH&E Management System

can be taken when unplanned-for events occur."[3] Tomorrow, one of your keenest competitors may buy out a less competitive company, and suddenly you are threatened with almost twice the competition that you have today. On the other hand, that same keen competitor may announce bankruptcy tomorrow, and you are challenged with potentially double the market that you have today. These are the types of contingencies that you should expect in the next five years. You cannot, of course, plan for each and every internal or external contingency, but generic guidelines that evaluate each issue can be written, so that people will know where to begin when the unexpected happens.

Consider the impact of the two competition scenarios mentioned above. If the first one happens, you may have to downsize, which is not a happy prospect. If the second one happens, you may have to ramp up rapidly to cover double the market before another company moves in on you. This is a happy prospect, but the pressure will be severe to win the race for available money and people, production capacity, reliable delivery, and the increased workloads on attendant functions such as the SH&E management system.

Contingency plans consider internal strengths and weaknesses and external opportunities and threats (strategic planners call this the SWOT process). A similar process occurred when you conducted the self-assessment and gap analysis. During these activities, the subcommittees determined where the strengths and weaknesses (the gaps) were and what opportunities or threats external stakeholders could present.

If and when OSHA issues a rule that mandates written safety and health programs, you will be presented with a threat or an opportunity, depending upon the effectiveness of the SH&E system in place. If you have effective safety and health processes with a low lost workday incidence rate, low absenteeism, high productivity, and a solid profit figure, you have an opportunity to stay on the course of continued improvement that you have planned and that is producing so well. It might also be timely for you to apply for VPP acceptance. Then, the OSHA Cooperative Consultation Office will be the program evaluators, not the OSHA Office of Compliance. With the Consultation Office, this is a cooperative activity, not adversarial, and it enhances the possibility of improved and constant worker protection.

If the SH&E management system does not meet these criteria, the threat or challenge is to play catch-up as fast as possible, and hope OSHA doesn't come to see you before you get it either started or completed.

Perhaps your thought may be, "We're too small. OSHA is not going to waste its time and resources on a small fry like this company." You may be right. Then again, you may not be. It depends upon where OSHA puts its emphasis for inspections. One year it may be health care facilities, steel mills, machine guarding, asbestos abatement, or fabric mills. Then again it could be office and industrial ergonomics. Company size is not a consideration. The emphasis is on the industry or operation. This is a contingency that is constantly present.

So what do you do?

All of the contingencies that may occur in the next five years cannot be foreseen. There are those that are the most likely, such as turnover, disability or death of a key staff member, a downturn or upturn in the market, or new OSHA rules. Then there are those that are the least likely, such as bankruptcy by the company or a competitor, or total destruction of the plant by fire, explosion, or sabotage. Trying to plan for all contingencies is, of course, not realistic. Planning for implementation of the new SH&E management systems is the first and most realistic priority.

Preparing contingency plans for least-likely events is similar to preparing an emergency response evacuation plan for the plant. It is more desirable that proactive measures are taken to prevent an emergency severe enough to activate the emergency response plan. But if it does happen, be prepared to initiate actions that will save lives and property. If there is no emergency response plan, the resulting disaster might resemble the aftermath of an explosion or a terrorist attack.

To prepare for contingencies, a three-point procedure is suggested by the *Applied Strategic Planning* authors:[4]

1. Identify scenarios for the least likely, but most important, internal and external threats/challenges.

2. Develop "trigger points" that will alert you to initiate actions for each contingency.

3. Agree on what actions to take for each of the trigger points.

For the purpose of this book's premise, these three points for contingency planning have been modified for the VPP-style of SH&E management systems.

Trigger points

Let's take as an example the issuance of the mandated written safety and health program rule by OSHA. This is a likely scenario because it seems certain that the mandate will be issued at some time in the future, and probably within the next five years. It will require certain responses from you. This is true especially if the company's safety and health performance does not meet the indicators that OSHA uses to evaluate program effectiveness.

Implement the Model SH&E Management System

What should be the trigger points or indicators that will initiate response to the issuance? There are two primary indicators that should be tracked carefully—one is internal and one is external. Internally, watch the lost workday incidence rate; externally, watch the company's industry rate. The OSHA log is one of the first documents that OSHA wants to see when they visit. OSHA uses the industry rate as an indicator of incident severity and frequency. Severity and frequency indicate hazardous work that causes multiple injuries. Hazardous work with an industry rate higher than the national average rate for all industries indicates to OSHA that maybe they should take a closer look at the companies in this industry. They want to learn what the hazards are and what the companies are doing or not doing to control them, so they target the companies, no matter the size, in that industry for special-emphasis inspections.

If the company's incidence rate is higher than the industry rate, it is likely that the company will be selected for an inspection. If the incidence rate is even close to the industry rate, the company may be selected. So, two of the trigger points are the internal incidence rate and the external national averages. If the rate is on an upward trend over a one- or two-year span, it is advisable to initiate an evaluation and analysis of SH&E programs for the root causes of the injuries and illnesses and begin corrective actions.

That is one set of trigger points that indicate that remedial action is necessary as soon as possible. Another set of indicators is to keep abreast of the OSHA and Congressional activities relating to new rules, OSHA reform, and issuances of notices of pending rules. OSHA and Congress talk openly about what their agendas and calendars are, what actions they have budgeted, and what priorities they have set.

Some of these activities are published in the Federal Register. Many of them are published in media that specialize in reporting what is going on in various fields, such as occupational safety and health, the Department of Transportation, the EPA, and legal rulings. Several organizations publish the occupational safety and health news, either in hard copy or on the internet. For example, OSHA's website provides the latest rulings, rules, and legal findings. If you belong to an industry association, they may also keep you posted on the latest Washington activities. See Appendix B, "Sources of Help."

For the SH&E management activities, there are certain internal and external indicators or trigger points that will spur the leadership to action. What and how much action is initiated will depend upon the particular state of the company and the effectiveness of its safety and health program.

Along with the safety and health performance indicators, pay attention to other indicators, such as workers' compensation costs, indirect costs relating to injuries and illnesses, productivity figures, marketing costs, and customer complaints. These, too, can be affected by absences due to injuries and illnesses, downtime to repair damaged equipment, and injuries serious enough to warrant OSHA inspections and citations. As discussed earlier, the impact of a poor safety and health program can be company-wide, affecting the health and safety of the employees, the costs involved in multiple injuries and illnesses, the productivity of operations, and finally the profit line at the bottom of the financial report.

Up to this point, the company's safety and health program may still be functioning as it was some months ago. Going through the preliminary steps to get to this point of the new process will probably take six to twelve months, if not more. How long depends upon the size and complexity of the company and the number of operational units or locations. Where official VPP participation is concerned, OSHA requires it to be site-specific. The size of the site doesn't matter—it can be 25 or 25,000 employees.

Executing the Strategic Plan

The strategic plan is simply a map of where the company wants to be five or ten years from now. In and of itself, it is an academic exercise. It helps to systematically organize and integrate the diverse activities of the company's operations that are required to improve, enhance, expand, and benefit from the revised SH&E management processes. It does not *do* anything. It is the guide to *get things done.*

The objective of this particular strategic plan is, of course, to actively get the SH&E processes and systems redesigned, organized, structured, functioning, and synchronized with the other essential parts of the organization. The ultimate outcome is that leadership and employees will benefit from all of the effort and time that has been put into the strategic planning. The challenge at this point is to ensure that the system is effectively implemented.

The authors of *Applied Strategic Planning* cite Tom Peters' (1984) definition of strategic management to mean that it "involves the execution of an explicit strategic plan that has captured the commitment of the people who must execute it; that is consistent with the values, beliefs, and culture of those people; and for which they have the required competency to execute." [5]

At this stage, the strategic management of implementing the redesigning of the SH&E program should meet Tom Peters' definition. You have a specific strategic plan that is tailored for the company to follow the requirements and criteria of the VPP. With a communication plan already in action, the employees' commitment should now be captured. A values scan and culture survey have been done and shaped the plan or the values and culture to fit the philosophy of the SH&E management system. Hopefully, a training program has also been initiated to prepare the employees to execute the planned approaches and activities.

Employee ownership

As pointed out several times in earlier chapters, employees must be involved. They must have a sense of owning the plan. Hammer and Stanton[6] discuss an informal survey that they conducted regarding the use of rental cars. In the survey, a number of people were asked if they had ever had a rental car washed at their own expense when obviously the car needed it. All of the answers were no. However, they had their own cars washed frequently or washed the cars themselves. The difference? Ownership. If, during the planning process, you have not involved the people who have to perform the implementing tasks—such as the machine operators, the warehouse people, the administrative support staff, the trainers, or the newsletter editor—an essential step has been omitted and another gap has been created. You have not provided for ownership by the employees of the new processes. That needs to be fixed before you take another step toward formal implementation.

Tangible Results

It is time to get to work on the real work that has tangible results—the formal accountability plan, baseline survey of safety and health conditions, analysis of hazards, and implementation of hazard preventions and controls, training, and related tasks.

Using the Critical Element Schedule (see Exhibit 5-3 as an example), activities are mapped into at least the next year. Detailed activities are sometimes better planned year to year, even though the overall plan allots three to five years for high-level areas. Changing conditions may make a course correction necessary at the end of the first twelve to eighteen months. Do not get too detailed too far ahead of anticipated results.

The current year's activities and tasks are outlined in the vertical and horizontal action plans discussed earlier. Start some safety and health-related activity in every department, operation, and function of the company. All departments and operations should witness that something positive and exciting is happening at all levels and in all segments of their areas. Observed activities means putting the accountability process into use, conducting the baseline survey, announcing the training programs schedule, and publicizing these efforts throughout the company. Plans are implemented concurrently by the employees responsible for them.

Tasks for the steering committee

The steering committee should generate the urgency, excitement, and enthusiasm for the implementation activities. They can also be the employees' monitors, supporters, and coaches. Leadership should be highly visible and articulate at this time, too—showing support, enthusiasm, encouragement, and being upbeat about the outcome.

Members of the steering committee may tend to get so involved in implementation that they begin to assume the management functions of the company. Be alert to this and continue to support and encourage them, but also assure them that the final decision-making authority must remain with the company's management team. Assign them the important task of continuous oversight of the overall implementation, holding regularly scheduled meetings to assess the progress, and submitting written progress reports to leadership of conclusions and recommendations for course corrections. Share the reports with the management team and the employees. In other words, the steering committee's job is to formulate the strategic plan; the management team's and the employees' job is to implement it.

Monitor the degree of implementation

To test the degree of implementation, monitor whether the employees (especially the managers) integrate the strategic planning into their routine safety and health duties or tasks. If the managers and employees have been oriented and trained to think of solving their safety and health problems according to the new direction, they will perform differently than they did under the old guidelines. Recall the incident discussed earlier about the hazard in the workplace and the employee saying, "No, I won't tell the supervisor so it can get fixed. I'll call maintenance myself to get it fixed." Such behavior is evidence that the redesigned SH&E processes are being implemented.

How long will implementation take?

For a small company that has all of its employees at one location, implementing a redesigned safety and health process will likely happen faster than it will at a large company with several hundred to several thousand employees, as well as two or three unions and a number of contractors. The strategic plan should reflect the complexities of integrating the redesigned processes into the structure and culture of the larger group. It should also reflect a longer time period to complete implementation.

At a small company, the planning team may be the management team and the subcommittees are the employees. The planners are also the doers.

Implement the Model SH&E Management System

The plan will have a shorter duration. How long will implementation take? The company will be in the implementing mode for as long as it adheres to the strategic planning philosophy and the VPP criteria. Both require continuous reevaluation, implementation, and improvement. The length of time for initial implementation activities will be at least until the first priorities are in place.

Getting Started

Schedule the date that activities will begin. Pre-planning, creating action plans, and setting priorities will help you get started. The strategic map is drawn up and you have identified what must be done, and in what order.

Since the subcommittees are already formed to handle the prioritized tasks within their critical elements, tasks can be done simultaneously in all of the elements. The priorities for the tasks identify when and where to start.

Perhaps you will start by training the machine shop operators. Theirs is hazardous work, so they get hazard recognition training first. Also, the baseline survey will begin in their department and immediate hazard controls will be initiated where a life-threatening condition is found. At the same time, meetings can be held with the human resources manager to decide how to formulate and implement the accountability system.

When implementation begins, also consider that this is a long-term effort. You may be putting into effect a five-year, or even a ten-year, plan. What is done today needs to fit tomorrow's environment or have flexibility built into it to readily respond to the changing nature of the company or the market. Very likely, what you do today *will not* fit tomorrow's conditions. Our business world changes fast, and leadership has to keep on their running shoes to stay up with it. Another beneficial attribute of the VPP criteria is that it leaves room to plan for tomorrow's changes by requiring evaluation and updating processes every year, or sooner if needed.

Create a stir

Leadership's job in the initial implementation stage is to begin as many tasks as feasible in each of the critical element areas, combined with involving employees and multifunctional work or planning teams. Create a stir, and make everyone talk to each other—many of the VPP criteria require that different functions talk to each other to reach agreement on the most effective course of action. There should be no more throwing tasks over the transom to each other. Keep the drum rolls sounding for interaction.

If a sequential workflow is projected, the plan will not be able to keep up with an environment that is on the fast track. Staying flexible and requiring your managers and employees to talk to each other and to other groups prepares all of you to respond readily to a change mandated by OSHA or by the market or the company.

This is especially true for sensitive areas such as finances; human resources; and safety, health, and environment rules. Every change in work environment creates either a small or large disruption in workflow, which can also cause unplanned, unexpected hazards resulting in injuries or illnesses and unplanned expenses. Anticipate such events, conduct hazard analyses, and institute proactive preventive measures.

Begin now, during implementation, to use cross-functional teams and multifunctional coordination and cooperation among all of your departments and operations. Spread the team effort that began with the multifunctional steering group for the strategic planning effort. The company will be well-positioned for the corporate structure of the future, which is predicted to be circular and cross-functional in structure. Initiate, as much as possible, systems that will serve as well five (or more) years in the future as they do today.

Celebrating and communicating the plan

Celebrate gains along the way. Completing the strategic plan and presenting it to the employees can be the first of many ceremonies to come. The authors of *Applied Strategic Planning* say, "The final announcement and presentation of the plan needs to be accomplished with the pomp and ceremony that signal an important event in the life of the organization. The effort in this rollout is to celebrate the creation of a plan with broad ownership—it is the organization's plan, not the planning team's plan!" [2]

This is the communications plan being implemented. The methods, techniques, and protocols of communication employed by marketing and media experts to convey a lasting message are discussed in Chapter 9. A common rule of thumb is to keep the message before the public, constantly and continuously. The communications expert at JSC encouraged constant repetition of the "JSC and VPP" message in various media by diverse messengers.

The activities of the VPP steering committee were regularly publicized in the JSC newsletter. The steering committee conducted workshops and seminars on the VPP management systems and processes, and arranged for employees, managers, and union representatives to attend the VPPPA national and regional conferences.

Implement the Model SH&E Management System

The Role of the Budget

In the case history of how Johnson Space Center got interested in VPP participation, OSHA found that, according to building codes, the hatchway entrances to the utility tunnels and building rooftops were too small by varying degrees. The significance of this violation is that the narrow opening would impede rapid egress in case of a life-threatening situation.

During the gap analysis by the worksite analysis subcommittee, repairing all of the hatchways at JSC was one of the major items on their list of gaps. To budget the work required the approval of several JSC offices. The job was estimated to cost $4–5 million.

The project was included in JSC's plant operations budget for completion over a two- to three-year period. This caused some differences of opinion among the safety, plant engineering, and budgeting staffs as to the criticality of the repair work compared to other work already planned for those budget cycles. In the 25 years of JSC's existence, there had been only minor injuries associated with the hatchways. Why was it so important?

Although unlikely to result in a serious event, the condition was one of high potential hazard. The high potential could not be ignored. OSHA obviously also felt that it could not be ignored, and the OSHA compliance safety and health officer issued a precaution for it. JSC management felt it was also prudent to agree with OSHA and chose not to postpone the work too far into the future.

As we are all aware, budgeting can have a severe impact on company-wide operations. This is one of the reasons that multifunctional teams are so important. Not only do the teams learn how each other functions and what priorities they have, they also learn to work together to solve differences before they become a disruptive reality.

When budgeting for the SH&E processes, be prepared for needs to be uncovered by the annual self-evaluations. As an example, JSC management did not foresee the hatchway problem. Although such surprises cannot be foreseen, the budget can include discretionary funds to deal with them. It is important to coordinate the budgeting and self-evaluation cycles so they can support the effectiveness of each other.

Accountability

The success of the SH&E *management* systems depends upon the *management* team being held accountable for carrying out their responsibilities. This, of course, also holds true for the employees. But accountability and performance must begin with the managers and supervisors before the employees believe this is truly a new system. In every element of the VPP, accountability by *all* employees is emphasized over and over. This book can do no less.

In the traditional management role of planning, organizing, and controlling an organization, executive leadership soon learns that an accountability system is necessary to obtain the expected results. This is also true of VPP-type management of SH&E systems. One of the first tasks is to include SH&E-related performance by managers and supervisors as an indicator of how successful and acceptable was the achievement of their job-related goals and objectives.

They in turn are responsible for holding their employees accountable for *their* performance. It is a cascading effect that reaches everyone. Incentive pay, bonuses, and promotions can or cannot be awarded to managers and supervisors depending upon their performance in implementing the new SH&E management tasks. If cooperation and implementation cannot be obtained any other way, squeezing wallets can result in astounding results.

In some VPP-participating companies, the employees are given personal safety and health goals and objectives. They are encouraged to report close calls and minor first aid incidents and are rewarded for doing so. They are recognized for suggesting improvements, investigating incidents, and other safety activities. The positive reinforcement of responsible and accountable safety and health performance contributes to achieving a safe and healthy workforce.

The JSC VPP coordinators developed specific performance standards for the JSC managers. The formal NASA appraisal form had these statements regarding health and safety:

> *Performance Planning:* ...Equal opportunity and health and safety must be considered in assessing performance for all managers and supervisors during each appraisal period...
>
> *Continuing Management Responsibilities:*....
>
> CONTROLLING
>
> - Health and Safety: Work environment is maintained in compliance with health and safety regulations. Health and safety hazards are promptly diagnosed and are remedied immediately. Periodic safety inspections are conducted. Action is taken to minimize instances of personal injury or unsafe use of material resources.

Using the above guidelines, the appraiser evaluates how the manager performed during the appraisal cycle. The guidelines were fine, but not specific enough for the safety and health processes and tasks. The suggested key specific objectives are shown in Exhibit 6-1.

The suggested specific objectives were distributed to all of the directors and managers by memorandum from the director of human resources on August 12, 1994. Excerpts from the memorandum read:

> ...Senior executive staff (SES) members, supervisors, and facility managers must place a greater emphasis on and be held accountable for institutional safety in their assigned areas. To ensure that this occurs...a safety element (will) be placed in the performance plan of each SES member, supervisor, and facility manager...Attached are specific examples designed for each managerial level that you may use or modify to meet your organizational needs...[7]

The performance standards and measurements reflect the VPP criteria requirements within the context of the JSC organizational structure. The ones that you develop and implement can be similar but worded to fit your company's accountability system and organizational structure such as the JSC document does.

Since governmental offices do not have bonuses, this type of reward was not mentioned. They do, however, have grade increases (with an accompanying pay increase) and outstanding employee recognition awards. Company leadership might want to emphasize that bonuses, pay increases, or pay-for-performance are directly related to meeting or exceeding VPP standards.

Incentive and Recognition Programs

Many companies use incentive or recognition programs as part of their safety and health motivational systems. Recognition programs seem to be gaining popularity over incentive programs. The fear among some employers is that cash bonuses or similar rewards for goal safety performance encourage underreporting incidents.

There are other aspects to cash bonus rewards. Take the example of a small company of about 155 employees who work about 350,000 hours a year maximum. Let's assume that their insurance experience modifier rate has been at or below 1.00 for a number of years. But when they experience even one recordable incident, their incidence rate is going to be above 1.00. Some companies will not allow contractors on their sites who have incidence rates above 1.00.

Other companies may have set a specific site-wide recordable goal for the year. If the recordable rate does not exceed the specific goal, the companies reward their safety manager with a cash bonus. The small companies who cannot guarantee a recordable rate below 1.00 are frequently not awarded con-

Exhibit 6-1. Johnson Space Center NASA Management and Supervisory Performance Appraisal, Proposed 1994

Performance Planning: Supervisor shall, to the extent of authority, furnish his/her employees employment and a place of employment that are free from recognized hazards that are causing or are likely to cause death or serious physical harm. Supervisor shall comply with occupational safety and health standards applicable to NASA with all rules, regulations, and orders issued by the Administrator of NASA with respect to NASA safety and health programs. (29 CFR 1960.9)

Performance Standards: (Johnson Space Center Management Directive [JSCMD] 1710.3G, "JSC Policy on Safety")

1. Issues and implements documentation necessary to comply with OSHA, NASA, and JSC safety policies and directives.
2. Implements appropriate protocols to review plans, procedures, and designs and to monitor operations within his/her organization for hazards to personnel or property.
3. Ensures that new or unforeseen hazardous operations or imminent dangers to personnel or property are shut down until risks are clearly understood by cognizant personnel and corrective actions are initiated and completed before operations resume.
4. Coordinates with SR&QA (safety, reliability and quality assurance directorate) Office during design, development, test, and operations on all hazardous flight hardware and ground support equipment. Ensures that hazards have been appropriately identified and adequately resolved by review of safety analyses, which may be verified by selected independent analyses by safety personnel.
5. Coordinates safety plans for potentially hazardous operations with SR&QA Office.
6. Ensures that mishaps occurring in his/her area are promptly reported, investigated, completely corrected, and lessons learned disseminated to his/her personnel.
7. Appoints safety representatives in accordance with JSCMD 1710.4.
8. Implements safety programs in accordance with requirements of applicable NASA programs, such as Space Shuttle, Space Station, payloads, etc.
9. Develops and maintains a safety plan describing how the safety requirements of JHB (JSC Handbook) 1700 are to be implemented throughout his/her respective office or area.

Key Specific Objectives:[*]

1. Ensures that employees are informed of JSC's safety and health programs and of the protection afforded employees through these programs.

(cont. p. 106)

Exhibit 6-1. (cont.)

2. Ensures that employees are aware of the location of the nearest medical treatment facility, correctly follow procedures for obtaining treatment, and report all occupational injuries or illnesses to their supervisors.
3. Ensures that employees immediately report hazardous conditions to their supervisors.
4. Takes appropriate action to protect employees in imminent danger situations.
5. Furnishes a safe and healthful place of employment and ensures that identified hazards are eliminated or controlled.
6. Ensures employees are knowledgeable of specific hazards associated with their workplace and duties.
7. Ensures their understanding and use of appropriate safeguards such as safety devices, caution and warning devices, and personal protective equipment; and trains employees in a manner that will ensure their safety and health.
8. Ensures that employees are informed of their specific responsibilities and rights under the Act (OSH Act), Executive Order 12196, and 29 CFR Part 1960 and how they may participate in the program.
9. Cooperates with and assists safety and health personnel while they are performing their duties as specified in the OSH program.

Specific Statements of the Measure of Performance:

1. ____ work area inspections were performed each quarter/during year.
2. ____ inspections were completed during appraisal period. Reports/checklists sent to ND4 (JSC Safety Office).
3. ____ mishaps were investigated and documented on Form 1627. Specific corrective actions are noted on Form 1627. Signs form.
4. ____ employees participated in the safety and health program activities. Activities were: List activities. Describe examples of employee involvement.
5. Attends safety and health professional development seminar, workshop, conference, training. Describe the activities.
6. Establishes goals and objectives. State goals and objectives for current appraisal period.
7. Had ____ mishaps current year. Had ____ mishaps last year.
8. Holds regular employee safety meetings. Sends copies of safety meeting minutes to ND4.
9. ____ new employees received orientation during appraisal period.
10. ____ employees received new or refresher hazard communication training during appraisal period.

* Critical element.

tracts, even though they may be the best qualified and most experienced. You can draw your own conclusions regarding the possible long-term consequences that could result from employing a less-qualified contractor to maintain and service the process lines or other critical operations in your company.

Stephen Brown's quote in Exhibit 6-2 ("We can't fix the problem if we don't know it exists") states the case for recognition, not incentive, programs. You can't fix problems with your operations until you know they exist. Employee involvement in the safety and health process takes the process of problem discovery and problem solving to the operating units and shop floors. Who else besides the worker doing the job knows the safest method of performing hazardous work? It is at the grassroots where you will find the greatest challenges and opportunities for shared authority, responsibility, and accountability. Not only does the VPP criteria require that safety and health problems be found before they become worse, employee involvement results in employee-driven solutions, implementation, compliance, and ownership.

Exhibit 6-2. Incentive Versus Recognition Awards

> I feel that an award based on workers not getting hurt inherently fosters non-compliance with reporting requirements.
>
> If an individual or group incentive is based on not getting injured, the pressure to hide incidents can be overwhelming. This can lead to improper medical treatment, a worsening condition, and repeat accidents because no incident investigation was performed to find the root cause.
>
> We (Potlatch) have gone to a recognition program based on employee involvement. Instead of paying people not to get hurt they can earn points through various safety activities. At the end of the year the more points you have the bigger the payoff.
>
> Points can be earned by running the monthly swift safety meeting or critiquing the same, reviewing or doing JSAs (Job Safety Analyses), performing lockout audits, doing hazardous housekeeping tours and safety observations. There is a multitude of ways to get the people on the floor involved in safety. This insures more safety awareness and involvement opportunities for everyone at the job site. It also does away with a major barrier to the reporting of accidents and incidents. As you all are aware we can't fix the problem if we don't know it exists.
>
> —Stephen Brown[8]

The Role of Corporate Leadership in Implementation

The active participation of corporate leadership in the implementation of the SH&E management systems cannot be overemphasized. Leadership leads the way by example, by their enthusiasm for the project, and by the messages sent, either written or spoken. It must be made clear to all employees that leadership is serious about the SH&E processes by the attentive reaction to an incident; by willingness to discuss the programs, anytime and anywhere; and by the programs that are put in place to improve the safety and healthfulness of the work environment. Many CEOs of VPP-participating companies and those that have VPP-type programs of excellence insist that *any* incident, no matter how minor, is reported to them within a specified number of hours after the occurrence, accompanied by an investigative report.

The CEOs of ten of the best-managed companies in America were asked in the late 1980s what they believed to be the critical elements of their job. Several of them agreed on: "(1) Setting the corporate strategy; and (2) Aligning the employees with it."[5] In any corporate-wide planning, these elements must be present for the plan to achieve its purpose. Leadership can only align the employees with the new safety and health processes by being aligned with them itself. They are the coaches, the mentors, and the boosters who maintain the momentum and morale, and recognize achievement at every opportunity.

Recognizing small gains

To maintain the momentum of enthusiasm and energy that has been aroused among the managers and employees, have a party each time that a milestone is reached, whether it is small or large. Celebrate. Put up banners, make it a front-page headline in the newsletter, distribute balloons, or have a picnic, just so the people who achieved the goal are recognized. Behavioral scientists found out years ago that one of our basic needs is recognition. A pat on the back is recognition in itself.

Recognition does not always need to be related to money or lavish gifts. Have a lunch catered in for the machine shop crew, with management present to comment on their achievement and good work. Their motivation and morale will go up to 100 percent. Try it—you will feel good yourself, and the employees will feel that what they do is important.

Keep this up. Don't just do it occasionally or randomly. Celebrate and publicize each short-term gain, especially in the early years of implementing the new SH&E systems. Consider the analogy of moving the Big Rock. You hand everyone a sledgehammer and they all start breaking pieces off the Big Rock and carrying them to the new location. Soon, it isn't a Big Rock anymore. In fact, it isn't even there. It is now small rocks that can be placed where they're needed.

Implementing the new SH&E management systems is the company's Big Rock. If everyone knows what his or her job is and how and when that job is supposed to be done, and they all fall to with energy and zeal, leadership will soon have an SH&E management system that yields heightened employee morale, lowered absenteeism, reduced injuries and illnesses, and improved productivity and profits.

Summary

This chapter has presented methods and means to implement the company's new SH&E management system. Up to this point, the steering committee has been in the planning and mapping mode. Covered in this chapter are preplanning for integrating action plans horizontally and vertically, and contingency planning that prepares for managing unexpected and unplanned events. The chapter discusses the activities and tasks necessary to install the new processes in daily company operations. The use of timeline schedules is encouraged to monitor completion of the action items, and suggests that monitoring be a steering committee task. Emphasis is placed on the corporate leadership's primary task of maintaining and demonstrating enthusiasm and support for the effort. The beginning of the implementation should be celebrated and publicized to raise awareness company-wide that this effort is special and worthy of notice. Budgeting and employee accountability are discussed as two major factors of successful implementation.

Now the company is entering the activities mentioned previously—plan-do-check-act. So far you have completed setting objectives and the plan/organize part. During implementation, you are in the "doing" stage. When the first self-evaluation is conducted after implementation, you will be in the "check" mode. When action is taken on what was found during the self-evaluation, you will be in the modification mode. Then, begin again at the planning stage to set objectives for the next cycle. The loop is closed.

References

1. Drucker, Peter F. *The Age of Discontinuity: Guidelines to our changing society.* New York: Harper & Row, 1969: 192–93
2. Goodstein, Leonard, Timothy Nolan, and William J. Pfeiffer. *Applied Strategic Planning: a comprehensive guide.* New York: McGraw-Hill, 1993: 283.

3. Ibid: 309.
4. Ibid: 311.
5. Ibid: 326.
6. Hammer, Michael, and Steven A. Stanton. *The Reengineering Revolution.* New York: Harper Collins Publishers, 1995.
7. Garner, Charlotte, JSC VPP files, 1993–94, published by permission of NASA-JSC.
8. Brown, Stephen, personal papers, published by permission of the Potlatch Corporation, Lewiston, Idaho.

7

Where Are You Now?
Annual Self-Evaluation of the Company's SH&E Management System

An effective self-audit procedure is part of a comprehensive safety and health program and should reduce employee injuries and illnesses, saving the employer costs resulting from absenteeism, workers' compensation and other insurance payments. An effective program may help reduce employee turnover and improve productivity. In terms of the OSH Act, the principal consequence of an effective audit program is a reduction in the number and severity of hazards, leading to a corresponding reduction in citations and penalties in the event of an inspection. A conscientious program should be particularly effective in eliminating high gravity, serious, willful, repeated, and failure to abate violations which carry by far the heaviest penalties...[1]
—Joseph Dear, Assistant Secretary of Labor (1993–1997)

The results of the SH&E management system evaluation enable you to fine-tune the system processes. The evaluation is a tool to assist you in ensuring the success and stability of your company. Paraphrasing OSHA's words, the comprehensive program audit evaluates the whole set of safety and health management means, methods, and processes, to ensure that they are adequate to protect the workers against the potential hazards at the specific worksite.

The self-evaluation is the one VPP program element that OSHA has found generally lacking or misunderstood by sites applying for VPP. That is unfortunate, because when your program evaluation is working well, your entire program is constantly improving. The self-evaluation provides both ends of the spectrum for the improvement process—where you are now, and where you want to be. This chapter offers one procedure by which you can conduct a comprehensive program audit.

Review the First Self-Assessment

You and your committee did a lot of work pulling together the information necessary to answer the questions in the assessment. These are relevant documents, especially the performance indicators with which you have been tracking program improvements.

Review the assessment documentation, the findings, the recommendations, and the completions. Also review the gap analysis documents to evaluate the closing (or not) of the identified gaps. In the evaluation you are identifying the accomplishments and non-accomplishments in each segment of the SH&E management system. During the assessment, it was mentioned that OSHA has a specific set of guidelines for the site approval evaluation team. These guidelines provide details of how to conduct a complete, all-inclusive self-evaluation. (See Appendix D, "OSHA VPP Site Evaluation Guidelines.")

OSHA VPP evaluators have found that there are three indispensable evaluation tools for judging the effectiveness of the SH&E management system:

- Review of written programs and records of activity;
- Interviews with employees at different levels;
- Review of site conditions.

Keep in mind that you are looking at the entire management system as one entity. The Shell Chemicals Deer Park Chemical Plant (DPCP) defines their health, safety, and environment (HSE) management system as:

> …a structured, documented set of activities designed to ensure and demonstrate that business objectives are met… . The purpose of the HSE Management System is to:
>
> - Use a structured approach to define HSE work processes which achieve and sustain compliance with HSE policies and applicable regulations.
>
> - Assure that HSE risks are managed effectively, enabling continuous improvement of HSE performance.
>
> - Ensure that HSE accountabilities at all levels are well defined, both within the DPCP Plant and between the Plant and other Shell organizations."[2]

Review of Written Programs and Records of Activities

Checking documentation is particularly useful for learning whether the tracking of hazard corrections to completion is effective. It is equally informative in ascertaining the quality of self-inspections, routine hazard analysis, incident investigations, and training. The documentation should indicate the degree to which accountability, disciplinary, and responsibility systems are applied.

Exhibit 7-1 is an extract from Section D, Appendix D, "Narrative Evaluation of Safety and Health Management System," *Suggested Format for Site's Annual Submission, Voluntary Protection Programs: Policies and Procedures Manual,* TED 8.4, effective March 25, 2003. This is a summary list of the elements that OSHA expects to find in place during a VPP-approval site evaluation by reviewing validated documentation, by interviews, or by site conditions. Some elements specifically mention "written," "documented," and "tracking." These words imply the necessity for written records to ensure that nothing falls through the cracks; these would include injury rates, management commitment and planning, and safety and health training. Comments, many of which have been extracted from material about each element furnished by Region VI, are included in the following pages.[3]

Along with reviewing each of these elements in your evaluation, you are also required by OSHA record-keeping rules to calculate the company's annual statistics regarding total incident rates, days away/restricted/transferred incident rates, and other statistics. These, too, become a part of the annual self-evaluation since they are another indicator of success or failure in the accident prevention efforts within the company.

Management Leadership and Employee Involvement

Management commitment. The assumption is that top management has committed to tailoring the company's SH&E system after the VPP criteria. Therefore, this section of the self-evaluation report will describe evidence of management's personal involvement. Particularly note in the self-evaluation the status of the company's top level SH&E policy. Are the employees knowledgeable about it, or do they display uncertainty about what it says and what it means? Record whether it is being taken seriously and, if not, ask of the system, "Why not?" Ensure that there is an explicit, clear description of the methods that management uses to display and

Exhibit 7-1. Summary of VPP Star Program Requirements

From OSHA Directive TED 8.4, Voluntary Protection Programs Policies and Procedures Manual, Appendix D, "Format for Annual Submission," p. D-4, effective March 25, 2003.

1. Management Leadership and Employee Involvement
   a. Management commitment to safety and health protection and to VPP participation
   b. Policy
   c. Goals, objectives, and planning
   d. Visible top management leadership
   e. Responsibility and authority
   f. Line accountability
   g. Resources
   h. Employee involvement
   i. Contract worker coverage
   j. Written safety and health management system
2. Worksite Analysis
   a. Hazard analysis of routine jobs, tasks, and processes
   b. Hazard analysis of significant changes, new processes, and non-routine tasks
      - Including pre-use analysis and new baselines
   c. Routine self-inspections
   d. Hazard reporting system for employees
   e. Industrial hygiene program
   f. Investigation of accidents and near misses
   g. Trend/pattern analysis
3. Hazard Prevention and Control
   a. Certified professional resources
   b. Hazard elimination and control methods
      - Engineering controls
      - Administrative controls
      - Work practice controls and hazard control programs
      - Safety and health rules and disciplinary system
      - Personal protective equipment
   c. Process safety management (if applicable)
   d. Occupational health care program
   e. Preventive/predictive maintenance
   f. Tracking of hazard correction
   g. Emergency preparedness
4. Safety and Health Training
   a. Managers
   b. Supervisors
   c. Employees
   d. Emergencies
   e. PPE

demonstrate top-level visible leadership in the SH&E system. Also cite examples of the activities that top-level managers and leaders engage in to clearly demonstrate their commitment. These could be chairing the management-level SH&E committee, always wearing the proper PPE when required, and regularly issuing personal safety messages to all employees.

Employee involvement. Describe the three (or more) activities of the SH&E process in which the employees are involved. Be specific about the nature and types of the employees' decisions regarding hazard assessments, hazard analyses, SH&E training, evaluation activities, or other SH&E areas particular to your company's business or operations.

OSHA does *not* classify the following as "meaningful involvement":

- Expecting employees to work safely;
- Wearing PPE;
- Participating on investigation teams, except as the injured or affected employee;
- Attending SH&E training;
- Other basic SH&E activities that are part of a regular program.

The number or percentage of employees should be stated. This provides you with an estimate of how many employees are involved. The same ones should not be involved constantly. *All* employees should be involved from time to time, and the evaluation assessment should clearly state this as an improvement goal (if involvement is not happening).

Goals and planning. List the company's goals and the objectives for reaching them. State how the SH&E system fits into the company's overall business cycle in regards to budgeting, resource allocation, training, and so on. Comment on how the employees were notified of the annual goals and objectives, how understanding of them was verified, and what methods were used to achieve them. If the employees were involved in developing the goals and objectives, describe how.

Organization. Describe how the company's SH&E system fits into the overall management organization. This can include a description of the lines of communication with management, reasonable access to top management, and a description of specific activities that include all employees, including contract workers, that are of equal high quality for company *and* contract workers.

Responsibility. Describe how the system applies the SH&E responsibilities for each employee. The evaluation report on this element should include a list of the job descriptions for all levels of employees, from managers to grassroots employees, and a statement that ensures there are no unassigned areas. Comment on the managers', supervisors', and employees' abilities to describe their SH&E responsibilities (as they pertain to their individual jobs) and the authority that they have been granted to correct deficiencies to safely perform the work. A specific example can be included to further emphasize how this element is being fulfilled.

Accountability. Describe the performance appraisal system for all levels of employees. Be specific—whether it is job performance evaluations, management by objectives, warning notices, contract language, or progressive penalties ending in termination for repeated violations.

Resources. Identify the SH&E resources that are available—certified safety professionals (CSPs), certified industrial hygienists (CIHs), environmental professionals, occupational nurses, and any other professionals who might be needed. Also, identify any internal or external resources that may be available, such as hazard monitors, fire safety personnel, SH&E instructors, and emergency responders.

Written safety and health program. Your written safety and health program should include all of the elements of the VPP criteria and state that they have been verified to meet OSHA requirements or industry best work practices. (If you are working at a federal worksite, also include the 29 CFR 1960 verification elements.)

Self-evaluation. Explain your procedure for conducting the annual SH&E evaluation. It must include all requirements of the VPP, an assessment of each requirement's effectiveness, specific recommendations for improvement, assignment by title or name of responsibility for completion of the recommendation, and target dates for completion. An assessment statement of the status and effectiveness of all the areas in Exhibit 7-1 should be included in the self-evaluation, whether it is an achievement or a challenge to improve. The assessment statements should be concise, specific, and describe either an improvement or a recommendation for further improvement, plus identification of the individual assigned the action, and the expected completion date.

Employee notification. Describe how the site notifies employees about participation in the SH&E management system and about their rights to register a complaint with OSHA and to obtain reports of inspections and accident investigations upon request. Some suggested means are bulletins, postings, toolbox meetings, group meetings, and e-mail (if available).

Contract workers. Describe the method used to ensure that SH&E working conditions are equal for all employees, whether contractor or company. Ensure that contractors are informed and understand that SH&E performance below the company criteria can cause termination of the contract and removal from the company site. Do not omit any contractor. The management system includes all work-related activities of the company or the contractor employees.

Worksite Analysis

Pre-use analysis. Explain how new/modified equipment, materials, processes, and non-routine tasks are assessed and analyzed for potential or existing hazards prior to use or implementation. Ensure that preliminary hazard analyses are conducted prior to integrating them into the company's SH&E system and the tasks involved.

Baseline hazard analysis. Describe methods used to record a baseline analysis to identify occupational health and environmental hazards associated with your specific work environment, such as air contaminants, noise, lead, and asbestos. Identify the SH&E professionals who were engaged in these activities. Be specific about the industrial hygiene surveys that concern sampling rationale and strategies. Your occupational health management system should include:

- Sampling strategies, including priorities of which are preventive along with exposures or spills;
- Initial screenings, and under what circumstances they are necessary;
- How sampling strategies cover all health hazards in the work environment and accurately assess employees' exposures, including duration, route, frequency, and number of exposed employees;
- How the sampling results are compared to OSHA permissible exposure limits to adequately assess employee exposure;
- The qualifications of the individual conducting the sampling and industrial health (IH) assessment;
- The system used to document and track the IH monitoring and the information included in the documentation.

Self-inspections. Describe procedures for conducting worksite SH&E inspections. Include specific information on:

- Hazard recognition training;
- Inspection schedules;
- IH sampling and monitoring;
- Who conducts the inspections;
- Tracking system to completion;

- Summarizing the testing and analysis procedures used and qualifications of those conducting them (where applicable to IH hazards).
- The OSHA VPP Revision in the Federal Register states the site must have:

   > ...a system for conducting, as appropriate, routine self-inspections which follow written procedures or guidance and which result in written reports of findings and tracking of hazards. In General Industry, these inspections must occur no less frequently than monthly and cover the whole worksite at least quarterly." [4]

   Be sure that the evaluation describes what you consider to be a comprehensive inspection and explains how you verify that you are performing monthly inspections and are covering the entire site at least quarterly.

- If you use a matrix of inspections to ensure all required inspections are conducted, describe the information contained in the matrix.

Hazard analysis for routine jobs, tasks, and processes. Describe the system used to examine and evaluate the SH&E hazards associated with routine tasks, jobs, processes, and/or phases. Describe examples and the forms used, if any. Identify the priorities, such as potential severity, historical evidence, perceived risks, complexity, and the frequency of jobs/tasks performed. In construction, emphasize special SH&E hazards of each craft and phase of work.

Identify the occasions when and under what circumstances you use safety or health professionals to assist with these types of hazard analysis. (Do not confuse this requirement with self-inspections covered under the preceding Self-Inspection section. What you are generating here are specific task hazard analyses and precautionary measures for the day-to-day jobs—installing pipe, opening/closing valves, assembling a scaffold, and similar tasks that have inherent hazards and particular precautions in operations, maintenance, administration, and other areas.)

Also include:

- How the results of these analyses are used to train employees to do their jobs safely;
- How the analyses contribute to planning and implementation of hazard correction and control programs;
- The analytical methods—job hazard analysis, job safety analysis, pre-job hazard assessment, monitoring, or other—used to identify the hazards at the site.

Employee hazard reporting system. Describe the written system that is available for employees to report hazardous conditions or practices. A mechanism must be in place for employees to report anonymously and there must be a mechanism to respond. Responses can be made via newsletters, postings on bulletin boards or e-mail, if available. (Sample hazard reporting forms may be attached to the self-evaluation document.)

Accident investigations. The procedure for investigating accidents, near misses, first-aid cases, and other incidents must be in writing. Describe:

- Investigator training;
- How you decide which accidents are investigated—at least the recordable injuries, illnesses, and near-miss incidents;
- When someone other than the supervisor conducts the investigation;
- How the root cause of the incident is derived (and that employee fault is not the only consideration);
- How the investigation results are used.

Pattern (trend) analysis. Describe the system used to analyze illness and injury trends over time by a review of injury/illness experience, hazards found during inspections, employee reports of hazards, and accident investigation. Include:

- How you collect and analyze data from all sources, such as injury/illness reports, near misses, first-aid cases, work order forms, incident investigations, inspections, and self-evaluations;
- What, if any, trending has been done for the last twelve months or longer;
- A description of the system to implement corrective actions after identification of the trends;
- The tracking system used to monitor corrections indicated by the trend.

Hazard Prevention and Control

Hazard elimination and control: engineering controls. Provide examples of engineering controls that eliminate or control hazards by reducing severity, likelihood of occurence, or both. Describe the methods for eliminating or controlling hazards, including:

- Reduction of pressure/amount of hazardous material;
- Substitution of less hazardous material;
- Reduction of noise produced;

- Fail-safe design;
- Leak before burst;
- Fault tolerance redundancy;
- Ergonomic changes;
- Guards, barriers, interlocks;
- Grounding and bonding;
- Pressure relief valves;
- Caution and warning devices, i.e., detectors and alarms in conjunction with the above.

Administrative controls. Describe how you limit daily exposure to hazards by adjusting work schedules or work tasks, such as job rotation.

Work practice controls. Identify the workplace rules, work practices, and procedures for specific operations to reduce employee exposure through:

- Changing work habits;
- Behavior-based training;
- Improving sanitation and hygiene practices;
- Other appropriate changes in how the job is performed. Include an example copy of the written safe and healthful work procedures for specific operations.

Personal protective equipment. Describe the written programs for the use of PPE, including respirators, hearing protection, ergonomic equipment, and other types of PPE used at the company site.

Hazard control program. Identify the major technical programs and regulations that pertain to your site, such as (but not limited to) hazard communication, hearing conservation, respiratory protection, lockout/tagout, confined space entry, and excavation and trenching. It is not necessary to include copies of these programs in your self-evaluation document, but you might want to identify the person or place in the company where these requirements can be found.

Safety and health rules/disciplinary system. Describe:

- The company's general safety and health rules;
- The written disciplinary system for enforcing the rules;
- How the system equitably enforces the disciplinary system for managers, supervisors, and employees;

Process safety management. For worksites subject to the process safety management (PSM) standard, describe the company's PSM system and assess the level of compliance against the standard's requirements, including:

- Employee participation;
- Process safety information;
- Process hazard analysis;
- Operating procedures;
- Training;
- Contractor;
- Pre-startup safety review;
- Mechanical integrity;
- Hot work permit;
- Management of change;
- Incident investigation;
- Emergency planning and response;
- Compliance audits.

Ergonomics. Describe the company's ergonomics program, including assessment of work areas and the system used to identify and track hazards.

Preventive/predictive maintenance. Describe the company's written system for monitoring and maintaining workplace equipment to predict and prevent equipment breakdowns. Include a summary of the types of equipment covered.

Occupational health care program. Include:

- Onsite and offsite health services and/or availability of qualified health care professionals;
- Coverage provided by employees trained in CPR and first aid, stating in which they are trained and if certifications are current;
- How employees on more than one shift have health care services available;
- How the company addresses hearing conservation and other health issues;

- How occupational health care professionals provide their services, i.e., design/implementation of health surveillance and monitoring programs.

Emergency procedures. Include a description of the company's emergency planning and preparedness program, emergency drills, and training schedules, including evacuations.

Hazard correction tracking. If you have not already included detailed descriptions of your hazard tracking systems in those sections that require tracking, include here the system for initiating and tracking hazard correction in a timely manner. Ensure that you have covered all identified hazards in or during:

- Accident/incident investigations;
- Self-inspections;
- Employee hazard reporting;
- Industrial hygiene surveys;
- Annual SH&E evaluation (prior year or years).

Describe how the items are tracked to ensure completion of abatement and the use of interim protective measures when necessary.

Safety and Health Training

Program description. Discuss the formal and informal SH&E training provided for managers, supervisors, and employees. Describe:

- The system used to ensure that all compliance and other required training is completed;
- How you verify the effectiveness of the training program.

Supervisors. Describe evidence that verifies supervisors:

- Understand their responsibilities;
- Meet their SH&E responsibilities effectively;
- Understand the hazards associated with the jobs performed by their employees;
- Understand their role in ensuring that their employees also understand and follow the rules and practices designed to protect them.

Employees. Describe:

- Records that verify employees understand the hazards associated with their jobs and the importance of following rules that apply to their jobs;

- How employees are made aware of the hazards associated with their work and in their work area;
- Methods used to teach employees to recognize hazardous conditions and signs and symptoms of work-related illnesses.

Emergencies. Describe the evidence that verifies that all supervisors, employees, contractors, and visitors know what to do in an emergency.

Personal protective equipment. Describe training or other methods to ensure employees understand:

- Why PPE is necessary;
- PPE limitations;
- How to maintain it;
- How to use it properly.

Interviews with Employees at Different Levels

Talking to randomly selected employees at all levels provides an indication of the quality of employee training and of employee perceptions of how well the program is running. If your communication program is successful and the training that you are providing is effective, the employees that you interview will be able to tell you about the hazards they work with and how they help to protect themselves and others by keeping the hazards controlled. Every employee should also be able to say precisely what he or she is expected to do during work and what they are to do if an emergency occurs.

Employee interviews can tell you about other aspects of your program as well. The employee's perception of how easy it is to report a hazard and get a timely and appropriate response will tell you a lot about how well your system for employee hazard reports is working. Employee perception of the system for enforcing safety and health rules and safe work practices can also be enlightening. If they perceive inconsistency or confusion, you will know that the system needs improvement.

For the information obtained by employee interviews to be useful, it is crucial that the employees trust the interviewers. It does no good for employees to tell you what they think you want to hear. They need to feel free to give you the bad news along with the good, without fear of reprisal. Confidentiality is a must. Be careful in selecting the people who will conduct the interviews. Select people who the employees trust and get along with, such as your safety person (or the person who performs safety duties if you don't have full-time safety) or someone in the personnel or human resources area. They may feel more comfortable with people who are outside of production and management or their particular work area.

Employee interviews are not limited to hourly employees. You can learn much by talking with first-line supervisors. Find out from line managers and your own senior staff what their perceptions are of their safety and health responsibilities. The mixture of responses—some positive, some negative, some uncertain—might surprise you.

Review of Site Conditions

Looking at workplace conditions reveals the hazards that are present. It can also provide information about the breakdown of management systems that were meant to prevent or control those hazards. If, in areas where personal protective equipment is required, you see large and understandable signs designating areas where PPE is required and employees in those areas, without exception, wearing their equipment properly, you have obtained a valuable clue as to how well the PPE program is working.

You can also obtain information from the hazards that are found through root cause analysis when contributing factors are identified and corrected or controlled. Such inquiry and analysis can identify inadequate safe work procedures or training, lack of enforcement of rules, no exercise of accountability, and confusion about the worksite objective of putting safety and health first. If an unsafe condition should be observed by your insurance inspector or an OSHA compliance officer, questions could be asked regarding whether a comprehensive survey of the worksite has been done by someone with enough expertise to recognize all of the potential and existing hazards.

These three methods of collecting data and information about your safety and health program can result in a thorough evaluation based on validated evidence. If you would like more information about the written self-evaluation document and how it can be structured, see Appendix B, "Sources of Help," and Appendix D, "OSHA VPP Evaluation Guidelines." The important points are that all of the elements are covered, that the strengths and weaknesses of each are discussed, and that you state what further actions you plan to correct the weaknesses and maintain the strengths.

Post-Audit Actions

In the Autumn 1992 issue of the *National News Report*[5] is an article entitled "Achieving and Maintaining Star Quality." Part VI (of six parts) concerns the safety and health program evaluation. The article notes that the annual self-evaluation "is the least understood and most misused of all the systems required for the VPP." The author states that "...the element that helps you know if you are slipping in any of these areas (the critical elements) is the annual self-evaluation."

Talk about the changes you found

Some participants use their annual evaluations to describe the elements of their safety and health programs instead of identifying what modifications or improvements they found during the evaluation. Sometimes, why the changes are needed is not discussed. Be sure that, when you and your steering group write the report of findings from the self-evaluation, you state clearly what improvement changes are needed and why.

The evaluation requirements state that the *effectiveness* of the safety and health systems and processes should be determined and the findings should be used to improve the implementation of the company's written safety and health program. The *National News Report* article notes, "it is the *effectiveness* of these program aspects that is to be determined by the evaluation." OSHA does not tell you in the criteria how to judge the effectiveness of your program.

In evaluating the process itself, ask, "Is the process producing the desired results?" If the answer is "no," go back and review the objectives to achieve during the evaluation that were set at the beginning of the VPP initiative. What purpose did you state for this new program? What purpose did you state for the evaluation to further the accomplishments of that ultimate purpose? What does your mission statement say? The mission statement should reflect your ultimate objective. Every action in the management of your safety and health systems and processes should be headed straight for the objective in your mission statement. Review the VPP evaluation guidelines again. Select your objectives for your evaluation to correlate with the guidelines.

If your steering group is missing the mark on evaluating the effectiveness of the systems they are examining, use the above technique to check how effective your system is to conduct your evaluation. We recommend reviewing your review methods, especially if this is your first evaluation. That's what the evaluation is for—to go to the root of the systems to uncover weaknesses or deficiencies.

OSHA allows employers to use "corporate or site officials or a private sector third party" to conduct the evaluation. Sometimes, if you and your senior managers conduct the evaluation, it may be lacking the objectivity to produce an impartial report of the conditions, and the resolutions may be unrealistically optimistic. New eyes see faulty or unsafe conditions that we may have been looking at for years.

On the other hand, we return to what the reengineering experts say about using consultants: Consultants have a knowledge of and experience with the VPP system, but no experience with your system. Consultants have an objective viewpoint and are committed and enthusiastic about the new system. Your people are the company's experts and have the corporate history. If you use consultants, you run the risk of transferring accountability for company processes to outside sources. There is a fine line here—the decision is yours.

Ongoing plan-do-check-act

Be ready to change, modify, include, or delete procedures in the system that are not generating the fulfillment of your objectives. These adjustments may be needed at almost any time in the life of a company, but they especially surface as a result of the annual evaluation. Whether the changes are in work environment, new technology, company growth, new OSHA rules, or ineffective systems or processes, your resolve will be tested to accept the challenge of change, to modify the old, or to develop new processes and systems that are required.

When Richard Holzapfel, directing the VPP effort at Johnson Space Center, read Hammer and Stanton's official definition of reengineering ("The fundamental rethinking and *radical redesign* of business *processes* to bring about *dramatic* improvements in performance"[6]), Holzapfel wrote in the margin of the first manuscript:

> For your information, this is a major component of evaluation—tossing out what doesn't work or contribute to company goals and objectives. Don't continue a program just because you've always done it. Think about Apollo 13 with NASA ground folks trying to be creative with only the resources on the spacecraft. You need to use imagination—look at things differently.[7]

A recommended process is to think of the system as planning it, doing it, checking or evaluating it, then acting upon the results—the plan-do-check-act circle. This is the ongoing nature of maintaining quality in any function, operation, or undertaking. It applies just as well to the SH&E management system.

At this juncture, it is important that you recognize that self-evaluation is not a "once-and-done" activity. It is an ongoing annual part of the SH&E management system. Just remember, however, that you do not have to get everything done in one time period. Action items can be scheduled and budgeted to fit into the overall company plans. What OSHA or any quality investigator looks for is the dedication, planning, effort, budget, and resources devoted to safety and health needs and requirements that are equal to those devoted to the stability and productivity in the other processes and systems in the company.

Benchmarking the Status of Your SH&E Management System

You have the VPP criteria to provide the elements and the intent. But what if you want to have some idea about what the general consensus may be regarding achieving excellence?

One of the better benchmarking tools you can use is issued by OSHA and is free to anyone who can access it on the Internet. This tool is the Program Evaluation Profile (PEP), which was issued in 1996 for OSHA compliance officers to use "in assessing employer safety and health programs in general industry workplaces." (See Appendix G.)

The PEP was cancelled on November 15, 1996, but is still provided on the OSHA website (www.osha.gov/SLTC/safetyhealth/pep.html) "as an EXAMPLE of an auditing tool to assist in the evaluation of a Safety and Health Program."

The PEP contains definitions of the levels of achievement in each element, including "excellence." It contains the critical questions that you want to ask of the results of your self-evaluation and a scoring method that you can use to derive a definitive status of your SH&E management system.

Summary

Chapter 7 discusses conducting the evaluation of your current safety and health management program to find out how it compares to the VPP criteria and what, where, and how critical are the unfinished action items from the self-assessment or the previous self-evaluation. The three methods used by OSHA to discover the site conditions and employee perceptions are included. You are advised to avoid the pitfall of describing how the program works instead of identifying what improvements have been achieved or what changes should be made to improve the program. We have referred you to the VPP Requirements, Evaluation Questions, and Evaluation Guidelines in the Appendix.

We cannot stress strongly enough the benefits that you will derive from an annual comprehensive self-evaluation of the company's SH&E management systems and processes. If your evaluation is as comprehensive as it should be, the resulting report will identify the achievements and the improvements still to be done, the quality (or lack thereof), related documentation, and the modifications or adjustments required to bring your SH&E management system up to the standard of the VPP criteria. You will be a believer by the time that you get into the second year of evaluating, setting new goals and objectives, making a concerted effort to put them into effect, and experiencing the improved performance, not only in SH&E indicators, but in operations, morale, absenteeism, turnover, productivity, and profitability.

Chapter 8 delves deeper into management commitment and employee involvement to further emphasize the importance of these two VPP criteria. The chapter offers particular recommendations on how to make these two elements happen effectively.

References

1. Dear, Joseph, Assistant Secretary, Occupational Safety & Health Administration. "Voluntary Safety and Health Audits Under the Occupational Safety and Health Act." September 11, 1996.
2. Shell Deer Park Chemical Plant. "Health, Safety and Environment Management System (HSE-MS), Administrative AD-22 CHEM." September 19, 2002: 3.
3. Klingbeil, William. "Information on VPP Element Approval Process." OSHA Region VI: Dallas, TX, November 22, 2002.
4. Occupational Safety & Health Administration, "Revisions to the Voluntary Protection Programs to Provide Safe and Healthful Conditions," Federal Register 65:45649–45663, 07/24/2000, "The Star Program," paragraph F5b3(a).
5. Voluntary Protection Programs Participants Association. "Achieving and Maintaining Star Quality," *National News Report,* Autumn 1992: 7, 27.
6. Hammer, Michael, and Steven A. Stanton. *The Reengineering Revolution,* New York: Harper Collins Publishers, 1995: 3.
7. Garner, Charlotte, and Patricia Horn. *How Smart Managers Improve Their Safety and Health Systems: Benchmarking with OSHA VPP Criteria.* Des Plaines, IL: American Society of Safety Engineers, 1999: 74.

8

The Two Linchpins

Leadership Commitment and Employee Involvement

That one can hire only a whole man rather than any part thereof explains why the improvement of human effectiveness in work is the greatest opportunity for improvement of performance and results. The human resource—the whole man—is, of all resources entrusted to man, the most productive, the most versatile, the most resourceful...
—Peter F. Drucker[1]

At a VPPPA conference workshop on employee involvement, a union representative who was a long-time company VPP representative related this episode:

> He was escorting a visiting union representative through the production area where he worked. They noticed an unsafe condition—a defect in the walkway, a broken railing, or something similar. The visitor remarked, "You better report that to the supervisor of this area so that it can get fixed." The union representative said, "No, I'm not going to report it to the supervisor. This is my work area, too. I'll call maintenance myself to get it fixed."

What did it take to create that attitude and the environment in which maintenance responds to an operating employee's problem call? It took leadership's willingness and commitment to trust the employees to do the right thing, to train them in what the right thing is, and to create the environment that makes the right thing easy and comfortable to do. It took leadership's willingness to convince the maintenance superintendent that any employee on the site had the right, the responsibility, and the authority to call the work order desk with a fix-it problem, and that maintenance had the responsibility to respond to the call and fix the problem. It took employees' willingness to accept responsibility for the parts of the safety and health program that they could do something about without fear of reprisal and to initiate the actions to get it done.

Employees' intimate knowledge of the jobs that they perform, the conditions in which they work, and the special concerns that they bring to the job provide a unique perspective that aids the success of the program. Pick up the phone and call maintenance or send an e-mail. That's it.

The workshop speaker went on to say that five years prior to the incident, the company was about to shut down that location because of the conflict between management and labor. Nothing was easy to get done, productivity was at a low point on the chart, and absenteeism and turnover were high. Then someone in management heard about how well the VPP criteria were working for the participating companies and decided to try the program at their site. It took five years, but the above episode shows that persistence does pay off.

To quote OSHA:

> Assignment of responsibility for safety and health protection to a single staff member or even a small group will leave other employees feeling that someone else is taking care of safety and health problems. *Everyone* in an organization has some responsibility for safety and health. A clear statement of that responsibility as it relates both to organizational goals and objectives and to the specific functions of individuals is essential. If *all* employees in an organization do not know what is expected of them, they are unlikely to perform as desired. Every employee must *know* his or her responsibility and *accept* the obligation to act.[2] [Emphasis author.]

The employees' acceptance of their individual responsibility and obligation to act depends upon leadership's commitment to and establishment of an environment of trust. The employee is certain of his right and responsibility because he trusts his leadership. Leadership trusts the employee's ability to carry out the responsibility in the right way. Although the final *decision-making* lies with the employer, OSHA and VPP participants have found that employee *participation* in decisions affecting their safety and health results in more effective management decisions, employee relations, and safety and health protection.

Stephen Brown, union VPP safety representative at Potlatch, said of management commitment and employment involvement: "The employees are the underused resource in the safety and health program. Management can dedicate the resources; that is, time and people. Give the people the time and they will do the job. Don't throw money at the problem. Allow the program to be employee-driven."[3]

Leadership Commitment

Some of the aspects of management leadership and commitment were discussed in a previous chapter, but there are a few more to consider as the company activates the refurbished SH&E processes. These relate to what the OSHA expectations are with regard to management compliance with the criteria.

It is leadership's responsibility and obligation to ensure unequivocal excellence and quality in this new approach to safety and health. Maybe about now, management is feeling overwhelmed with the multiple commitments, processes, and systems that they are being assigned. However, there is experienced help available for the price of a telephone call. The OSHA Cooperative Consultation Office regularly observes and evaluates employers who are implementing these criteria. Through your local or regional OSHA office, their consultants are available to answer questions and to assist you. And they will not write a citation if they visit your operations unless they observe an imminent life-threatening hazard and you decline to take immediate remedial action.

If management has not been through the challenge of giving up control of an important function of the company operations to the employees, this could be their personal trial by fire. The systems' criteria require it. OSHA says employee involvement means involvement in meaningful activities that have a real, measurable, and decisive effect on the outcome and intent. "Meaningful activities" are presented in Exhibit 8-1.

The point now is management's willingness to surrender the reins of the SH&E processes to the employees who must put the processes into practice. Why is that necessary? One of the VPP expectations is that ownership of the employee parts of the processes is transferred to the employees. Managers are forming a partnership with employees to share all

Exhibit 8-1. Meaningful Employee Involvement in SH&E[4]

> OSHA's VPP requirements list specific suggestions regarding "meaningful employee involvement:"
>
> **Employee participation in general industry.** Employers must have at least three ways that employees are meaningfully involved in the site's safety and health problem identification and resolution.
>
> 1. Appropriate and acceptable means of meaningful participation methods as listed in the "VPP: Policies and Procedures Manual" are:
>   - Safety committees
>   - Audits and self-inspections
>   - Accident/incident investigations
>   - Safety and health training of other employees
>   - Analyses of job hazards
>   - Suggestion programs and planning

elements of the SH&E processes. Management must change top-down from being the controller to being the leader, coach, and mentor, reserving only the right to make final decisions. The employees become the movers and shakers.

When you are faced with these decisions to let go and share your power with others, consider Stephen Covey's discussion of the *abundance* and the *scarcity* mentalities. The *abundance mentality* is one with "…a bone-deep belief that 'there are enough natural and human resources to realize my dream' and that 'my success does not necessarily mean failure for others, just as their success does not preclude my own.'" The *scarcity mentality*…"tends to see everything in terms of 'win-lose.' They believe 'There is only so much; and if someone else has it, that means there will be less for me.'" [5]

When you practice abundance thinking, you realize that by sharing the tasks, responsibility, and accountability with the people who are working alongside you, you have added their strengths and weaknesses to yours to create a consolidated whole. Like single strands of manila rope wrapped into one stronger coil, your strengths will compensate for their weaknesses; and their strengths will compensate for your weaknesses. A stronger, more reliable, more resilient organization results. You have a company in a win-win condition.

When the employees feel the same level of responsibility for their own safety and health that leadership feels for them, you will have a workforce that has accepted accountability for their own actions—to themselves, to you, and to the company. Until full employee acceptance occurs, nothing changes—leadership is vulnerable to OSHA's compliance actions, your employees are vulnerable to their work environment, and the company is vulnerable to an unknown injury frequency and severity. You as a leader must be willing to be less so that you can be more by an extension of your responsibility, authority, and obligations to the employees.

There is a price that must be paid for the win-win condition. The price is not easy for most corporate leaders; that is, leadership must give up control of some of the company affairs to the employees. If not surrendered completely, then control must be shared. The literature is unanimous that this is one of the most difficult, if not *the* most difficult task for the executive to perform. After all the years that executives spend planning, organizing, controlling, and managing the company and their careers, they must change their methods of internal control in order to attain the superiority and excellence of performance that is described here.

The internal control that we describe makes managers become less managers and more leaders, and makes the employees become more managers and controllers. Making yourself become less a manager but gaining stature as a leader results in gaining more from the employees. The final result is an outstanding net gain for the executive, the employees, and the company.

Leaders must learn the skill of being open to the influence of others when corporate-wide decisions are to be made. Hierarchy has no part in the win-win status that we are discussing. It does not matter that you are the CEO, the top manager, or the superintendent. What does matter is that leaders are so committed to the welfare of their employees and their company that they are willing to involve all employees in the running of the business, keeping them updated on financial status, listening to them, acting upon their recommendations, and recognizing their abilities and untapped capabilities. Giving up control is not easy. "Easy" and "success" are seldom a matched pair.

The personal commitments that OSHA expects of leadership are stated in detail in Appendix D, "Format for Annual Submissions," TED 8.4, *Voluntary Protection Programs (VPP): Policies and Procedures Manual*. A brief review of them appears in Exhibit 8-2, and some remarks about a few of them follow.

If yours is a voluntary effort not to be monitored by OSHA, staunch and unswerving dedication to the SH&E process goals and objectives are nonetheless essential if you want to be successful.

More Leadership Personal Commitments

Union agreement

Even though the company is voluntarily complying with the criteria, you still must have the acceptance of all unions on the premises if you want the program to succeed. Senior managers and union representatives must participate together to transform this redesigned initiative into a successful reality. Union participation is absolutely essential. Without it, your efforts are doomed.

After getting the steering committee familiar with the VPP, the first meeting that the JSC VPP coordinator held was with the president of the JSC civil service union and the union representatives working for the JSC contractors. He invited Patricia Horn, (co-founder of the OSHA-VPP criteria) to the meeting to explain the meaning, purpose, and scope of the VPP and how the programs concerned and involved union employees. It so happened that two of the representatives already had union workers in VPP participant companies. They praised the programs highly and strongly supported JSC's intention to qualify for acceptance—a lucky shot for the JSC coordinator and the initiative.

Exhibit 8-2. Leadership Personal Commitments Required by OSHA VPP Criteria[6]

Policy and goal. This concerns your written occupational SH&E policy and your current visionary goal for the safety and health program and the annual objectives for meeting that goal. Be sure that the policy and goal are concisely written. The policy can be short or long, depending upon your personal style.

Labor and union agreement. If you are applying for official VPP acceptance, you are required to include with the application a signed statement by the authorized union agent that the union either supports VPP application or that they have no objection to the company participating. If the statement is not included, or will not be, OSHA will deny the application. If the union does not support the initiative, OSHA recognizes that the union can severely impede the accomplishment of the goal and objectives.

Written assurances for actions that OSHA has determined are critical for VPP success.

Commit to:

1. Doing your best to provide outstanding safety and health protection to the employees through management systems and employee involvement.
2. Achieving and maintaining Star Program requirements and to the goals and objectives of the Voluntary Protection Programs. (The Star Program requirements are the ones covered in this book.)
3. Correcting, in a timely manner, all hazards identified through self-inspections, employee reports, accident investigations, or any other means with interim protection provided as necessary.
4. Protecting from discriminatory actions (including unofficial harassment) of any employee with SH&E-related duties.
5. Providing the results of self-inspections and accident investigations to the employees upon request.
6. If yours is a construction company, recording together injuries for all employees at the site, no matter who the employer is, and basing the calculation of your injury and illness rates on the consolidated numbers.
7. Maintaining the appropriate documentation relating to your safety and health programs.
8. Preparing a written annual self-evaluation document and annual safety and health statistics.
9. Notifying employees of their rights under OSHA rules and describing the methods used to ensure that all employees, including new hires, are notified.

Leadership is also expected to include these specific activities in your safety and health systems and processes.

Management planning. How safety and health planning is integrated into comprehensive management planning.

(cont. p. 135)

Exhibit 8-2. (cont.)

> ***Written safety and health program.*** Include the critical elements in the SH&E program, i.e., management leadership and employee involvement, worksite hazard analysis, hazard prevention and control, and safety and health training and how they are being implemented. Also include the above written assurances.
>
> ***Visible management leadership.*** How leadership provides visible leadership in implementing the safety and health processes, and how the function fits into the overall management organization. Leadership and commitment are enhanced if the SH&E functions report directly to leadership.
>
> ***Employee involvement.*** How the employees are involved in the safety and health processes and what are the specific employee decision processes by which they affect the critical elements of the safety and health program. See Exhibit 8-1 for specific "meaningful activities" to involve employees.
>
> ***Contract workers.*** What methods are used to ensure that safe and healthful working conditions are consistent for all employees, even where more than one employer has employees at the same site, such as contract workers as well as full-time employees.
>
> ***Responsibility, authority and resources.*** How authority is granted to enable assigned responsibilities to be met and how resources are obtained and used.
>
> ***Line accountability.*** Include leadership's system for holding line managers and supervisors accountable based on some type of evaluation of supervisors.

Line accountability

One of the leadership obligations is to link responsibility, accountability, and authority to opportunity. Show the employees how they can achieve safer and more healthful working conditions by their involvement and their acceptance of the opportunity, the responsibility, and the accountability, along with authority. They also will experience fewer injuries and illnesses and achieve increased productivity in their jobs and excellence and quality in the safety and health processes.

These are a few acceptable types of accountability systems. Other types may be equally effective and can also be acceptable. The key is that all managers know they are being evaluated on the effectiveness of the way they carry out their SH&E responsibilities.

The accountability process may be:

- A performance rating system that includes safety and health;
- Management by objectives (MBO) SH&E goals;

- A system of rewards for safety and health performance and a disciplinary system for managers whose employees do not perform their work in a safe and healthful manner;
- A system involving a central safety and health committee consisting of top managers and chaired by the plant manager with impact down to the first-line supervisor, so long as evaluation of performance of SH&E responsibilities is implicit.

There is more discussion about accountability in Chapter 6.

These are actions and activities that only corporate leadership can initiate and perpetuate. Whether the company is small, medium, or large, there are ways and means to implement the processes successfully. The OSHA VPP criteria expect results. How the process is implemented is leadership's decision. Sophisticated and expensive methods are nice, but not necessary.

Max De Pree concludes his book, *Leadership Is An Art*, with: "Leadership is much more an art, a belief, a condition of the heart, than a set of things to do. The visible signs of artful leadership are expressed, ultimately, in its practice."[7]

Writing it down is not enough. You must do it, and everyone at your facility must be able to recognize that you are, in reality, leading.

Employee Involvement

Employee involvement provides the means through which workers develop and express their own commitment to safety and health protection for themselves and for their fellow workers. Employee involvement and participation is a common thread that runs through successful leadership texts. The emphasis is always upon establishing ownership by the employees. What we are involved in, we own. It is part of us because we invested that part in it. If we are not involved in making a change happen, we react with "that's *his* (whoever "he" might be—usually management's) idea, not mine. Why should I do that (whatever "that" is)? It wasn't *my* idea, let *him* do it." Stephen Covey says, "Involvement is the key to implementing change and increasing commitment. We tend to be more interested in our own ideas than in those of others."[8]

Resistance, not acceptance, usually results from imposed change. Including the insight and energy of all employees in the processes and systems facilitates the achievement of the goal and objectives. Management provides the encouragement, the parameters, the boundaries, and the direction. Together they develop the means and the methods for employee participation in the structure and operation of the SH&E program and in the decisions that structure the program.

What are some of these methods and means?

The Mobil Joliet safety and health committee distributed a flyer at the 1996 VPPPA Conference. It said in part:

> Employees are involved in the Safety, Health and Environmental Committee, incident/injury investigations, job hazard reviews, work area safety audits, safety and housekeeping audits, safety policies and procedures development, as well as other aspects of the safety program.

More "methods and means," as described in OSHA's VPP version in the Federal Register are summarized in Exhibit 8-3. This is taken from an earlier version of the VPP Guidelines. It includes examples of other meaningful activities and may be useful for construction companies.

Challenges and opportunities abound in the involvement of employees in these processes and systems. As discussed earlier, leadership is challenged to develop the *"abundance mentality,"* which says, "I am not going to lose anything if I share with you—there is plenty for all and to spare." Through employee involvement, leadership is providing space for the creative and innovative potential of the employees to emerge. At the same time, the leaders are providing a vision for a better, continuously improving future. The space and the vision can inspire and motivate the creativity of those who will help you achieve the goals and who will benefit the most from the results.

Involvement means that leaders and their employees are sitting at the same table discussing a common problem. Employees contribute to the solution and carry out their share of the tasks to resolve the matter. It is now as much the employees' problem as it is management's! During the resolution process, they have had a chance to air their reservations or negative opinions, but only in the positive direction of improvement. Resistance and complaints have been diffused. Progress toward positive results is advanced.

What do employees say about this reengineered or redesigned safety and health program? The episode related at the opening of this chapter exemplifies the results that can be obtained—the employees recognize hazards and take the necessary steps to correct them.

Stephen Brown, the VPP union safety representative for Potlatch Corporation Consumer Products Division in Lewiston, Idaho, spoke to a safety and health conference held in Houston, Texas, on March 18, 1997. Here are some of his remarks about Potlatch's participation in the OSHA programs:

Exhibit 8-3. OSHA-VPP Requirements for Construction Site Joint Labor-Management Committee for Safety and Health[9]

1. Committee members must have a minimum of one year's experience providing safety and health advice and making periodic site inspections.

2. Committee must have at least equal representation by bona fide worker representatives who work at the site and who are selected, elected, or approved by a duly authorized collective bargaining organization. A union representative who is not a company or subcontractor employee cannot be a committee member. Employee representation must be at least equal to management; more employees' members are acceptable but not more management members. OSHA also expects the management members to work onsite.

3. On occasion, OSHA might accept as ex officio members of the committee higher-level company and union officials who are not always onsite, provided that the full committee functions in accordance with VPP requirements and is not merely pro forma. (This kind of arrangement might be particularly applicable to the construction industry, where many union locals may be represented overall by the Building and Construction Trades. In such cases, OSHA expects higher-level officials to provide continuity in the implementation. Usually, however, these high-level officials still serve more effectively in an oversight role, rather than as actual committee members.)

4. In construction, trades that will be onsite for the bulk of the project are the best choices for employee members. The committee composition may be changed, however, as new trades come onsite, either by increasing the committee size or maintaining the size and replacing trades who have fewer workers onsite.

5. Committee meets regularly, at least monthly, keeps minutes of the meetings, and follows quorum requirements consisting of at least half of the members of the committees, with representatives of both employees and management required. The minutes will include actions taken, recommendations made, and members in attendance.

6. Committee meetings should include activities such as review and discussions of committee inspection results, accident investigations, safety and health complaints, the OSHA Log, and analysis of any apparent injury trends.

7. Committee makes regular inspections (with at least one worker representative) at least monthly and more frequently as needed, and has provided for at least monthly coverage of the whole worksite.

8. Joint committee must be allowed to:
   - Observe or assist in the investigation and documentation of major accidents.

(cont. p. 139)

Employee Involvement

Exhibit 8-3. (cont.)

- Have access to all relevant safety and health information; such as the OSHA log, workers' compensation records, accident investigation reports, accident statistics, safety and health complaints, industrial hygiene survey and sampling results, and training records.

- Have adequate training so that the committee can recognize hazards, with continued training as needed. The test is whether the committee members possess adequate experience and knowledge to conduct inspections, or whether training is needed for them to adequately recognize hazards. The committee may need training in other areas, such as how to use statistics to direct inspections, and how to work together as a group. There should be some mechanism for determining training needs and for providing these needs.

In late 1993, Potlatch called all of their employees in to announce we needed to downsize to remain competitive in the forest products industry.

When the cuts began in 1994, morale, safety, quality, and production nose-dived.

With less resources from the company to work with, we had to find a way to utilize what we had to protect our membership and jobs, considering that we felt safe mills would be the ones that stayed open.

Through the VPP process, we found a way to empower an underused resource—our own workforce. Through the VPPPA mentoring program and the networking we learned how to broaden our employee involvement to the point that when OSHA did our onsite (evaluation), they stated that they had never seen a better employee involvement program.

Brown then went on to describe the details of Potlatch's employee involvement program, which is outlined in Exhibit 8-4.

At the VPPPA OSHA Region VI Chapter Conference in Dallas/Fort Worth, Texas, March 12, 1997, Manuel A. Mederos, Director of the International Brotherhood of Electrical Workers (IBEW) safety and health department, spoke on the Voluntary Protection Programs. Some of his remarks are cited below:

> Our endorsement of these programs (VPP) is based upon our belief that an employee is the company's most precious asset.

Exhibit 8-4. Speak Out: Employee Involvement[10]

Remarks by Stephen Brown, VPP Union Representative for Potlatch Corporation, Lewiston, Idaho, to a conference on safety and health, Houston, Texas, March 18, 1997.

Meaningful employee participation is a basic element of the Voluntary Protection Programs. The Guidelines emphasize the balance necessary between management and hourly employees' participation in the site's safety and health program by linking these two components. At Potlatch's Consumer Products Division in Lewiston, Idaho, we use a variety of ways to attain true employee involvement at all levels of our safety and health program. They include hourly safety hierarchy; safety observers; near-miss and hazard notification; team groups or problem solvers; training of other employees; doing site inspections; participating in incident investigation; writing job safety analyses and process hazard review; and safety and health committees.

The hourly safety hierarchy begins with each department's shift safety committee person. This person assists the shift supervisor in safety matters, runs the monthly safety meeting, and helps with incident investigations that occur on his/her crew. The second level consists of area safety chiefs who work with their department superintendents and shift committee people to address safety concerns in their area. Finally, we have four union safety representatives who coordinate all hourly safety involvement. Kent Lang, one of our union safety representatives, also serves as our site VPP Facilitator.

We use hourly safety observers in a process we call POWER (Process of Workers Eliminating Risks). POWER is led by two full time union facilitators and a 14-member steering committee composed completely of hourly workers. They have trained nearly 100 employees to do voluntary and confidential observations to gather information to help create a proactive safety environment. Management is only asked to provide the time and resources needed to make POWER a success.

Near-miss and hazard notification is accomplished by a safety work request procedure. When an employee has a safety concern he/she fills out a safety work request form and turns it into their shift supervisor. If the concern cannot be immediately addressed, it goes to the department safety coordinator to be tracked to completion. The status of any safety work request can be checked anytime by calling up the tracking log.

The Guarding Team is a prime example of how we solve problems by using teamwork. We have had many machinery guarding issues, including design effectiveness and priority assignment. By using a team of hourly technicians, machine operators, and maintenance workers, a successful guarding program has been initiated.

Employee training is accomplished in a variety of ways. New hires are orientated by the union safety representatives. Job specific training is done by experienced workers and documented. OSHA required training is done by the four department safety coordinators, two of which are hourly employees. Safety committee people are trained by their area chiefs, all graduates of the OSHA 501 course.

(cont. p. 141)

Employee Involvement

Exhibit 8-4. (cont.)

Safety inspections are done on a frequent schedule. Employees inspect their equipment, tools, and job site daily. Hazardous housekeeping tours are done monthly and the union safety representatives participate in the VPP quarterly site inspections.

Employees are involved at all levels of incident investigations. Whenever an accident, incident, or near-miss occurs, the initial investigation is conducted with the input of the shift safety committee person. The area chief is included so that all shifts can be informed on what has occurred. Finally, the union safety representative signs off when satisfied that the root cause has been identified and corrective measures are in place.

All JSAs and PHRs are employee-generated at Potlatch's Consumer Products Division. We feel that employees know their jobs and how to do them the safest possible way. JSAs are incorporated into our employee training.

Most worksites have some sort of central safety committee that functions as a recommending body on safety issues. Ours consists of 12 members, half of which are hourly employees. Chairmanship of this committee rotates yearly between management and union members. Subcommittees with specific concerns, such as ergonomics or lockout/tagout, are generated from central safety. Sign-up sheets are posted and employees with expertise or concerns in these areas volunteer to work. This has proven to be a very successful way to attack various safety problems.

Finally, we have a Corporate Environmental, Health and Safety Hotline. Employees can call this 800 number 24 hours a day to report any safety concerns they may have. Messages are checked daily by a corporate attorney and all concerns are given immediate attention. This service maybe used anonymously or if the employees leave their names, progress reports will be given to them.

VPP sites know the value of true employee participation and are proud of their involvement. We are at Potlatch's Consumer Products Division in Lewiston, Idaho.

Text of Mr. Brown's presentation reprinted by permission of the Voluntary Protection Programs Participants' Association

Any program which acknowledges and taps the expertise and dignity of the individual worker deserves our attention.

Partnership programs have had problems as they progress. You may be coming from a relationship where there has not been a mutual trust, an adversarial relationship had existed, or one where management really had a problem sharing the responsibility to manage with its employees. Sometimes the problems are insurmountable. However, sometimes the parties fail because it is easier to fail than work out their differences. Once they work out their differences they find out that it wasn't so bad and, lo and behold, they also started to find trust in one another.

A look at the IBEW finds that we have 18 local unions involved in VPP at ten companies representing over 27,000 employees at 16 sites…Nine of the ten companies have achieved Star status at this time.

I did a quick survey of our local unions, and I did not receive one critical comment from any of them. Believe me, if they had one I would not have to survey them to find out that there was a problem. Some of their comments were:

'There is a stronger commitment to safety by our IBEW members, due to the opportunity to be more involved with the making of safety procedures, and our input is welcomed by management.'

'After our first year, we are very pleased with the direction we are headed, but we have a lot to do.'

'The VPP gives us the employee involvement and management commitment necessary to have a good safety and health process.'

'Severity of injuries has decreased, worker compensation rates have decreased, and the reporting of injuries has increased (more minor injuries have been reported) due to employee trust in the program.'

'One last thing, the reduction in injury rates and savings on workman compensation costs is a great measure of your work. However, the real value you perform because of your work is sending your fellow employees home safely.' [11]

How to Get This Done

"This is all well and good," leaders may be thinking, "but how am I going to get all of this done? My plate is full now and stays full. Where am I going to get the time?" We cannot advise on where you will get the time, but in our experience, people make time for what they believe is important. As this program gets underway, you will find your time being spent in different ways.

Suggestions are offered about how to get this done. The strategic plan should include schedules and priorities for getting the activities in motion. Setting up safety committees, safety observation programs, ad hoc safety and health problem solving groups, and committees to plan and conduct SH&E awareness take pre-planning. Company size and how small and middle-size companies can make all of this work has not been mentioned. These how-to's and who's are covered in Chapter 6 on implementation and Chapter 10 on continuing improvement.

Employee involvement in the SH&E processes takes discovering and solving problems to the operating units and the shop floors. It is at the grassroots where you will find the greatest challenges and opportunities for sharing authority, responsibility, and accountability. Employee involvement results in employee-derived solutions, implementation, and compliance.

Summary

VPP leadership commitment and employee involvement requirements are summarized in this chapter. One of the most useful elements of the quality, empowerment, and reengineering concepts of the VPP criteria is employee involvement. Even though it is one of the most productive and beneficial actions that the corporate leadership can take, it is also one of the most difficult—the release of operational control to the employees. It seems a paradox: To assert internal control over the management system, management must yield control of the operational processes and programs.

But that is the way this initiative works best. Leaders draw the map, plot the course, and study the business conditions, the obstacles, and the rocks. The employees follow the map and travel the course. Stephen Covey talks about running alongside your employees. Leaders are there to assist and support, to direct the travel, to point the way past the obstacles and the rocks, to coach and encourage, and to have resources ready. Leaders and the employees become a team to accomplish the goals and objectives of the company. Involving employees in the decision-making processes of the company will create a sense of ownership on their part. If they have a part in shaping the structure of their tasks and are responsible and accountable for performing to high standards, their faith and trust in leadership and the company will be heightened and solidified. Leadership commitment and employee involvement are truly the linchpins that serve to hold together the elements of this complex SH&E management system.

In previous chapters, the criteria and the requirements of the OSHA-approved VPP safety and health management processes and systems have been discussed. Chapter 9 covers putting the plans in action—at last.

References

1. Drucker, Peter F. *The Practice of Management*. New York: Harper & Row, 1954: 262–63.

2. OSHA Office of Cooperative Programs. "Voluntary Protection Programs to Supplement Enforcement and to Provide Safe and Healthful Working Conditions," *Federal Register*, 53, no. 133, (July 12, 1988) as annotated by OSHA Instruction TED 8.1a, May 24, 1996.
3. Brown, Stephen. "Employee Involvement," *The National News Report* [now *The Leader*] Summer 1996.
4. OSHA Instruction Directive 8.4, "Voluntary Protection Programs (VPP): Policies and Procedures Manual," March 25, 2003. Chapter 3, paragraph II.C.1.b.
5. Covey, Stephen R. *Principle-Centered Leadership*. New York: Summit Books, 1991: 157–58.
6. OSHA Instruction Directive 8.4, "Voluntary Protection Programs (VPP): Policies and Procedures Manual," March 25, 2003. Chapter 3, paragraph C.1.
7. De Pree, Max. *Leadership Is An Art*. New York: Doubleday, 1989: 136.
8. Covey, Stephen R. *Principle-Centered Leadership*: 217.
9. OSHA Office of Cooperative Programs. "Voluntary Protection Programs to Supplement Enforcement and to Provide Safe and Healthful Working Conditions."
10. Brown, Stephen. "Employee Involvement."
11. Mederos, Manuel A. Remarks before VPPPA Region VI Chapter Conference, March 12, 1997, published in *The Leader* (Summer 1997): 40–43.

9

The Stakeholders
Culture and Communications

For the enterprise is a community of human beings. Its performance is the performance of human beings. And a human community must be founded on common beliefs, must symbolize its cohesion in common principles. —Peter F. Drucker[1]

After sixteen years of successfully selling fire protection and safety engineering services to the petrochemical industry and to NASA at the Johnson Space Center, the management of a small company, based in Houston, Texas, decided to take their corporate culture to a higher level.

In 1990, inspired by Tom Peters' study, *In Pursuit of Excellence,* and the fourteen points of Dr. W. Edwards Deming, they held several leadership conferences of the company's corporate and senior leaders and technical staff. Although this was an intensive and expensive effort for the company, the two owners say today that they wouldn't have done any differently. When one of the co-owners was asked how long it was before they began to see changing behaviors and attitudes, he replied that it was four or five years—but the results were worth it. Through the leadership conferences, which emphasized Deming's fourteen points of quality, the supervisors and technicians learned to contribute to the solid customer relationships that the company had built. The field personnel are the company's primary representatives with their customers. The owners attribute a great part of the customer trust that the company enjoys to the positive attitudes and superior work of their field personnel.

As a result of their studies of quality and excellence, early in their pursuit of excellence they published vision, mission, standards, values, guiding principles, and continual improvement statements. Their stated mission is:

> To continually improve the economic well-being, quality of life, and future of all of our stakeholders, including our internal customers (employees), our external customers, our suppliers, and others.

Their guiding principles are:

- Guiding Principle Number 1: Respect the worth and dignity of every human being within the company and elsewhere.
- Guiding Principle Number 2: Be trustworthy in every word and action.

One of the co-owners says to remember these are only words on a piece of paper. They are not even worth reading unless they are practiced every day. They form the foundation for any program or process that the company institutes to provide continually improving service to internal and external customers. Where operational and safety and health processes are concerned, the company leadership look to management systems to identify root causes and controls, just as Dr. Deming's fourteen points and the VPP criteria do.

Changing the Company Culture

What do values have to do with modifying the SH&E management processes and systems? And why do you need to evaluate the company culture? Because the management processes and systems that are being proposed in this book are different from the traditional or generic type of safety and health program plan. This is not a structured, hierarchical process, nor a closed process that only the top management and the safety and health staff develop, implement, and administer. These processes join with every other process and operational system in the company. Every employee is involved or it doesn't work. It is leadership's task to tell the employees why the changes are necessary in such a way that they not only understand, but they buy in to it and *want* to be a part of it.

No matter the outcry of doubt or the fearful anxieties that may result from the first announcement, leadership and the steering committee must continue to look after the best interests of the company. There is a greater need here—to control the corporate destiny where OSHA safety and health practices, procedures, and costs are concerned. Reform of the OSH Act, the OSHA mandate for a written safety and health program, and an assessment of performance measured against OSHA rules are getting closer and closer to reality in Washington, as OSHA continues its effort to become more of a partner than an enforcer.

As an example:

In November 1997, OSHA issued OSHA Instruction CPL 2-0.119, which implemented OSHA's high injury/illness rate targeting system and cooperative compliance program (CCP). In OSHA's words, the CCP is:

> ...an alternative enforcement strategy which offers some employers a choice between a traditional inspection and working cooperatively with OSHA to reduce injuries and illnesses in the workplace.

OSHA randomly selected 80,000 employers with 60 or more employees in manufacturing and other industries. From the 80,000, OSHA collected site-specific injury and illness data. They developed an inspection program consisting of three categories:

1. The *high injury/illness rate targeted inspections* included the employers from the initial list (with a lost workday injury/illness rate of 7.0 and above) and those who agreed to participate in the CCP.

2. The *nonresponder inspections* included the employers who did not respond to the data collection request and consisted of a records inspection.

3. The *records verification inspections* included 250 randomly-selected establishments to evaluate the accuracy of the information submitted by the employers in response to the data collection request. According to OSHA, "OSHA's visits to these establishments will start with an analysis of injury and illness records. Further action may be taken if the reported data is inaccurate."

OSHA continued oversight of these companies for five years. To measure success of the CCP, OSHA conducted annual program evaluations of the CCP participants. OSHA collected annual information from the OSHA 200 (now 300) records of the CCP employers during the five years. Additionally, OSHA continues to collect annual runs of data from the CCP-participating and nonparticipating establishments under other emphasis programs. This approach allows OSHA to look for a reduction in an individual employer's lost workday injury/illness rates, as well as to determine any reductions in rates across individual industry groups.

Readers are encouraged to access the latest site-specific targeting (SST) directive for additional information of how OSHA selects the industries for the SST list, and which industries [by Standard Industrial Classification (SIC), or North American Industry Classification System (NAICS)] are targeted.

If your company is in a high-hazard industry group, urgency mounts to begin the initiative immediately. Action is especially necessary if your safety and health program is less than acceptable or if the employees are experiencing injuries serious enough to result in days away from work or an entry on the OSHA log. Such circumstances make it even more urgent to be ahead of these targeted, programmed inspection mandates by OSHA. Establish voluntary safety and health processes by following criteria that are *already approved by OSHA*. You will ensure safe and healthful working conditions for the employees now and into the future, with the added advantage of enjoying a cooperative relationship with OSHA.

But these processes not only require changes in the management practices of safety and health systems, they also require a change in the attitudes and behaviors of the employees. Corporate leadership is obligated to explore and clarify the corporate culture, values, and beliefs that are shared and exercised every workday. These also shape the values and beliefs followed away from work. Corporate leadership should ask itself: "Are we proud of the values and beliefs that our employees take home with them every day?" Do the values exercised in the company culture serve as behavioral standards for staff, managers, and employees in whatever environment they may be—work, home, or community? Corporate culture, mission, values, and beliefs are the framework on which any employee, operation, function, or process in the company can rely to provide measurements against which to assess personal choices and decisions, performance, productivity, and profitability.

How do we start?

Take the first step to initiate the safety and health processes and systems as required by the OSHA VPP. Leadership is required to develop a work environment that encourages employees to actively participate in the SH&E process. Clarify the corporate mission, values, and beliefs, since they must be able to support the involvement of everyone in the company.

Values and beliefs were discussed in previous chapters. Readers were cautioned that the safety and health values must be in harmony with the corporate culture. It is again recommended that management closely examine corporate values, beliefs, and mission to ensure that they can embrace a new set of values and beliefs.

Go through the values survey process, similar to what was done at the Johnson Space Center. Find out what the managers and the employees believe the company mission and values are. Are the values cooperative, tolerant, understanding, and generous? Does your company have an "all for one and one for all" work environment? If not, there is some work to do before the VPP requirements can be met.

Refer to the bibliography for more sources of information on value scans and guidelines for establishing a values-based SH&E program management process.

Once the values survey process is completed and analyzed, the steering committee will know how close together corporate culture and the VPP SH&E safety and health values are. The next step is to train the managers and employees in these new values and how they fit into the company culture so they are applied every day in the workplace. Communicate these new values, what they are, and how they will benefit everyone in the company. As stated in the chapter introduction, that is just what the small company did.

Then, establish guidelines based on the safety and health and company values by which you can evaluate implementation and performance. The guiding principles quoted at the beginning of the chapter are an example: "Respect the worth and dignity of every human being within the company and elsewhere. Be trustworthy in every word and action."

Look for changes in behaviors, attitudes, and actions where safety, health, and environmental work practices are concerned. Leadership, managers, and employees will have to internalize these new ways of doing SH&E and develop a new attitude toward the requirements—assuming responsibility, taking control, and getting things done. Start the orientation with leadership, supervisors, and senior staff. This becomes the core group to carry the message of the new goal to all of the other employees.

Keep in mind that employees *want to do their best. They want to stand out,* to be recognized and rewarded; to be important. Since the 1960s and 1970s, behavioral scientists have stated that recognition and reward is a basic human need. But to perform their best, to be worthy of recognition and reward, employees must have a work environment that trusts them to do the right thing, that encourages them to exert their creativity and to stretch their abilities, that is tolerant if they make a mistake, and that gives them the information they need to perform to everyone's benefit. Leadership and senior staff are the ones to make this happen.

Leadership makes this happen by ensuring that they set a high standard for the values that guide them and the employees personally, and for the values that guide the company.

Develop company values that recognize people values

Start with people values first, the ones that your company employees will be inspired to follow. The values of trust, truthfulness, and trustworthiness are the most significant among the authors of value-based systems.

Others that are held in high regard are being honest, empathetic, sympathetic, straightforward and open, unselfish in concern for others, and generous with time and knowledge to help others learn. These traits and habits are needed not only for today's SH&E safety and health management systems, but for the corporate structure of the future. Most of us have these values taught to us as children. If leadership, staff, and employees are not practicing them, now is the time to begin._

Then look at the company values. It is easy to identify where they may not be synchronized with the people values. The company values should have an impact on every task and job within the company, and should influence the success of accomplishing tasks and jobs.

You should measure how well the values benefit the company. Ensure there are guidelines to control them, to change their direction, and to eliminate them when they no longer serve the company's best interests. The test of the usefulness of the company values is to evaluate whether the outcome from following them has a positive and good effect on the people, the organization, the community, or the environment.

Think of a few of the ways the work environment has changed in the last thirty years. Thirty years ago, the majority of employees did not know how the chemicals they were working with could harm them. Today they must be informed and protected. Thirty years ago we didn't realize, as the human race, how we were destroying some of our valued natural resources just by the destructiveness of our habits. Now we are keenly conscious of how fragile our environment is. In many cases, the values that guided us thirty years ago are not valid today. Ask yourself, "Is this value still good for_____?" and you fill in the blank. If you do not foresee a productive, positive, beneficial outcome, select a value that will be more lasting.

The company values should support the people values if leadership wants people to support the company. Business values must be tailored to the business and to the uniqueness of the products, and must provide an umbrella for the other missions and values developed by the operations, functions, and departments, including the company-wide SH&E processes and systems.

If the company's established mission and values and the values required for VPP SH&E processes are not synchronized, revise the established mission to incorporate the new values. If employees already accept the premise that change is constant, a revised mission should not cause severe vibrations in the emotional environment of the company. This may not even be the first time that the company mission has altered course. Most assuredly, as the company travels down the road of continuous improvement and company growth, it will not be the last time.

Accepting responsibility

When the mission, values, and beliefs are synchronized, one of the next challenges will be how to transfer the duties and tasks of the new processes to the managers and the employees, and how to get them to accept the responsibility for the duties and tasks that have heretofore been "safety's job." You cannot just dump these tasks on their desks and say, "It's all yours, ladies and gentlemen," walk away, and think the results will meet expectations. Managers and employees will need formal training in how to accept responsibility. After the training, managers will be called upon to exercise mentoring and coaching skills for some time to come. This is when they start running alongside of their employees, as Stephen Covey says, as coach, mentor, and resource supplier.

Probably the group with the highest resistance level will be the middle managers and supervisors. Whenever there is a corporate change or shift, they feel the most insecure, threatened, and vulnerable. Their comfortable niches or mini-empires are about to be invaded. Their first instinct is to defend their turf.

The first formal training program initiated at the Johnson Space Center was for all of the facility managers. As the JSC opinion survey showed, when it came to actively supporting the safety and health program, the authority link between top management and grassroots employees was the weakest at the middle management connection. They simply did not (and some still do not) accept as part of their jobs the administration of the safety and health tasks in their areas. Unfortunately, this is common.

There are a few techniques you can use to help your managers and employees accept their responsibilities for these new tasks. Fortunately, the VPP requirements are in your favor with regards to responsibility. One of the foundation stones of the criteria is that *everyone* in the company is responsible for their safety and the safety of others. From the very beginning of the program, everyone must be involved and stay involved. Involvement can begin with training and educating *everyone* about what his or her responsibilities are, how they will be affected, and what benefits they will experience from assuming their safety and health tasks.

To carry out their responsibilities and involvement adequately, the employees must recognize that priorities must be set and schedules must be kept to complete the job. Protecting themselves from life-threatening situations must take priority over meeting production quotas. A scheduled, structured set of job steps must be carried out before a confined space entry job, before climbing a scaffold, or before lifting a box of files.

Leadership contributes to and facilitates this process by clearing the way for everyone to accept his or her assigned responsibility. Remember, once again, the incident where the employee said to his visitor, "No, I won't report that to a supervisor. This is my workplace. I'll get it fixed myself." It took management actions to create the environment so that the employee *could* get it fixed. Clear the obstacles that may block the lower echelons from exercising their responsibilities to perform their safety and health tasks. Make it easy for them to develop ownership.

To solve employees' problems relating to their safety and health responsibilities, encourage them to be creative about solutions and to think in terms of how the entire company might benefit from their ideas. They know more about the actual doing of the job than managers and leadership. Their ideas may be something that you would never think of, but they could be effective all over the company. Allow employees to exercise self-expression and to determine the outcome of their jobs. Encourage them to talk to managers and leadership about their ideas for improving the job. Convince them that they have the right to express themselves, even if they are taking a chance that the idea may not be suitable or feasible. You never know until you hear it. Once it is brought up, it can be discussed and accepted or rejected.

Encourage them to not be afraid to take the chance. Rejection is not failure, and it does not mean they are wrong. It just means that the particular idea doesn't fit the situation at this time. There are always options for another course of action. Rejection is simply a temporary condition that can be resolved another way. Let them finish the job once they solve the problem. Let them go through all of the steps of making it work.

Once the mission, values, and beliefs are in harmony with one another, the time is right to communicate the purpose and scope of the new SH&E processes. The formal safety and health training, the briefings, and the orientation meetings were the beginning of the communication process. There are other aspects of a communication plan that will help you.

The Open Window of Communications

Max De Pree says about communication, "There may be no single thing more important in our efforts to achieve meaningful work and fulfilling relationships than to learn and practice the art of communication."[2] He may be referring to the individual's communication skills; but they apply as well to company communication practices.

The "communication plan" has been mentioned several times in previous chapters. No matter how small or how large, a company cannot survive without communication among individual employees, managers, departments, customers, and suppliers.

Communication of all types goes on constantly among people—it may be good or poor, positive or negative, constructive or destructive. It is just as continuous among company employees, no matter what their jobs or titles are with the company. It is frequently inaccurate, negative, and destructive. In fact, this kind seems to travel the fastest through the company grapevine even when distance separates the communicators. Means of communication are so accessible in our electronic world that it is just too easy to pass on juicy gossip.

The methods and means of communication are the circuits that connect all in the company. You, personally and collectively, cannot function without it. You want to prevent, circumvent, or defuse the negative talk. You want your messages to be effectively delivered; and you want them to be accurate, complete, informative, and friendly.

Effective communication has certain attributes that should be applied to the company's communication plan. In every reference in this book, you will find the admonition to "communicate, communicate, communicate." Communicate about the VPP SH&E management systems, their purposes, and their benefits. Communicate about the company business and how well it is doing or how and where it needs shoring up. Communicate goals, objectives, mission, and values. In other words, communicate about all of the factors that constitute the company characteristics.

In addition, certain guidelines should be followed to communicate effectively. Remember that personal behavior, attitude, and examples are one of the means of communication. What you do or say, the employee observers will also do and say. Recall the guiding principle cited at the beginning of the chapter, "Respect the worth and dignity of every human being within the company and elsewhere." Let your communications convey your respect for the employees' worth and knowledge. A message that is conveyed truthfully, courteously, and respectfully is a message received positively and responsively.

Remember that the employees have a right to know what hazards they are working with, how the hazards will affect them, and how they can protect themselves. If you were in their shoes, you would want to know. They have a right to know what management is doing to protect them and any other information that is relevant to their work. State your messages simply and clearly, and ensure that your managers do, too. Max DePree says that good communication is "based on logic, compassion, and sound reasoning."[3] Base your communication plan on these fundamental guidelines and you will have a plan that will create teamwork and improved job and safety performance.

After the strategic plan was in place, the VPP coordinator at Johnson Space Center began communication planning. He said, "Communication is the most important part of launching this new program." He publicized the positives of the program, even the small ones.

The communication plan

Communication is the key to success in any company-wide initiative, no matter what its message. This applies just as well to launching the effort to achieve an efficient and effective safety and health program. One of the most important tasks that the corporate leadership can accomplish in implementing the VPP criteria is to implement a communications program for managers, supervisors, and workers.

Behavioral psychologists say that change starts with communication. Change begins with the implementation of the criteria. Paradigms shift, and the company culture and environment is affected. The good news is that the change is in the positive direction. Skillful and thorough communications will ease the path to achieving goals and objectives.

Marketing experts use a number of techniques that will encourage positive employee reaction to the new ways of work. As described in the following paragraphs, these experts also recommend particular media and the amount of information to provide for the different groups of employees. We have already discussed a few of the techniques—putting the steering group together, starting with the senior staff and then expanding to a vertical group representing all levels in the company; defining the program's mission, values, and beliefs; and constructing the strategic plan.

Creating a sense of urgency to implement the changes is a technique that belongs in particular to top management. Urgency is motivating and stimulating. It implies a challenge and a possible threat to an individual's status quo. It arouses people's defensive mechanisms and excites them to action.

It is for leadership to articulate the reasons why the company not only urgently needs these systems, but why they are crucial to the company's survival as a profitable competitor. Another urgent issue concerns the plans of OSHA compliance officials to mandate safety and health programs and to assess them according to their own program evaluation profile rules. The response to this urgency is to propose the proactive approach of implementing voluntarily a set of criteria already approved by OSHA, thereby setting up beneficial barriers to avert the OSHA-forced compliance and monitoring. Of course, the primary urgency is to adopt zero injuries and illnesses as the company's ultimate goal. The sooner actions begin to prevent injuries and illnesses; the sooner zero will be reached.

Other reasons for urgency are that:

- Almost immediately, the employees will experience the benefits of the continuously improving working conditions, which will be a visibly positive result of the new systems.
- Over the long term, the company will benefit economically by reducing the direct and indirect costs of safety and health and by increasing productivity and profitability.

In a volatile and changing competitive marketplace, any internal change that can reduce costs and increase productivity should be integrated immediately into the company operations. To create the momentum of urgency, prepare to kick off the communication plan as soon as the talk begins about reasons for change.

The following are practices and methods successfully used by consumer goods marketers.[4]

Communication methods and practices

Divide into similar groups. Divide employees into similar groups with similar jobs, such as the senior staff, the marketing department, human resources, shop workers, the first-line supervisors, the departmental middle managers, and production superintendents. Tailor the messages to each group, because each one has different attitudes, behaviors, needs, and tasks. Timing, media, and emphases must be different for each group. To develop their messages and select media, marketers ask questions such as these about each group:

- Who is in this group?
- How will the redesigned SH&E management processes affect them?
- What will their reaction be to the new way of doing SH&E?
- What behavior is needed from the group for the new systems to be successful?
- What messages should they hear to realign their behavior consistent with the new processes?
- When is the best time to deliver the messages?
- What medium should we use to deliver the message?
- Who is the appropriate messenger for this group?

It is important to be specific and target each individual group with a message that fits its functions and place in the organization.

JSC's VPP coordinator drew a large wheel with spokes and inner wheels to visualize his concept of the sectioned communication plan. Figure 9-1 represents the functional segments of the target audience. The groups sharing the same circle going around the wheel require the same fundamental message, varied to fit their segment of the workplace. The smaller Figure 9-2 fits into the hub of the larger wheel and represents the safety and health program elements to be communicated to the various segments.

The VPP coordinator then identified each segment as a group—for instance, space and life sciences; plant operations; and safety, reliability, and quality control—to indicate that each segment has particular safety and health conditions. Inside the wheel he identified levels of employees—

Exhibit 9-1. JSC's Segments of Communication

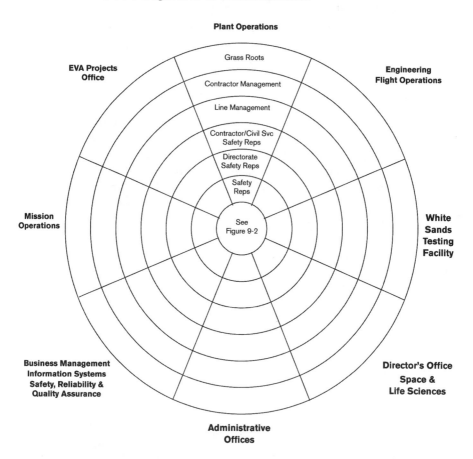

Source: Johnson Space Center, NASA, Institutional Safety Office

civil service and contractor safety staffs at the hub, directorate safety representatives, contractor and civil service safety representatives (with corollary duty), line management, contractor management, and grassroots—signifying that the employees within the segments require varying deliveries of the message.

You might consider this the vertical and horizontal dimensions of the communication strategy. Sometimes, JSC prepared messages just for one level of employee in one directorate (department), such as the line managers in the engineering directorate. Sometimes a grassroots message was prepared that went to everyone. Many times the messages were transmitted simultaneously. What was communicated depended upon the needs of the audience.

Exhibit 9-2. JSC's Safety Management Office and Safety Elements Relating to Communication

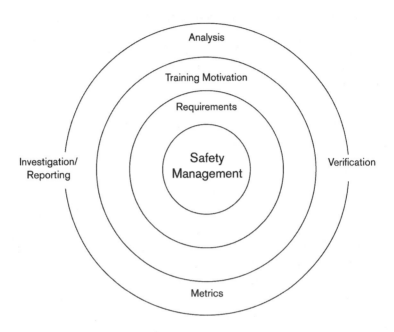

Source: Johnson Space Center, NASA, Institutional Safety Office

Use all available methods of communication. Media should be tailored to each group. Tailor delivery of the message to each target audience. For instance, leadership's message to the senior staff will be different from the message to the shop workers. The senior team may have a roundtable discussion. With the shop workers, an all-hands or departmental group meeting will be called.

Use different media for the type of information to be conveyed, remembering that direct and personal communication is the most effective. On the other hand, it may be appropriate to broadcast the message to all employees at one time. In this instance, use the company newsletter, special bulletins, the company computer network, closed-circuit television, or whatever available media can reach all on the same day and approximately the same time.

To discuss the new processes and systems with the middle managers and the supervisors, review the eight questions in *Divide into similar groups,* above, and address their needs appropriately. The same technique can be used to present the message to the shop foremen, the union representatives, and the technician employees.

Use a variety of voices. People soon tire of hearing the same message said the same way by the same person. A major soft drink company has a firm policy to change the delivery or the medium of their message on a regular frequency no matter how successful a campaign may be. Change to a new delivery of the same message before the employees get tired of the first one.

The same principle applies to getting people to buy into the SH&E processes. Shop foremen will perceive and convey the message from a different perspective than the superintendent; the plant manager can brief everyone on recent or planned activities regarding the initiative; leadership can update everyone through the newsletter or by a personally signed memorandum to each employee. Each message should discuss the new effort and the information in his or her individual style, which will enhance the message.

Keep the message simple, direct, and specific. Use everyday language to communicate the message. Don't use the SH&E technical jargon; use words and descriptions that everyone can understand. Be specific. Describe as precisely as you can why the SH&E systems are going to change, what parts are going to change, and how the change is going to happen. Most importantly, talk about the benefits each change will bring to those listening or reading the message. For example, you might say something like:

"We are going to raise the level of how we conduct our safety, health, and environmental activities to a higher plane. Why do we want to do this? Because:

- "We are experiencing injuries and illnesses.
- "We need to reduce or prevent those incidents as soon as possible.
- "There is an OSHA program that we propose to use that has proven results in reducing injuries and illnesses, in establishing a cooperative relationship with OSHA, and in benefiting the company financially.

"We believe that this positive and proactive approach to SH&E processes will result in cooperation with OSHA because we are voluntarily instituting a management system that OSHA has already approved. We believe also that it will help us to reduce or prevent injuries and illnesses and to achieve a constantly improving and more productive safety and health work environment for all of us.

"So, what's going to change?"

Then go on to cite specific examples of the systems *tailored to the company operations* that will be implemented—in the formal SH&E training, worksite hazard analysis, hazard identification and control, and employee involvement—and how the elements will be implemented in their departments or operating units.

For example, the requirements include responsibility and accountability. You may have some employees who have probably never thought about being responsible for safety or being accountable for either a negative or positive outcome resulting from their actions. The message to each segment of the workforce should describe as precisely as possible what this means to them. To senior managers, it means possible high marks or black marks on performance appraisals, or perhaps even termination for more serious events. To shop workers it means complying with the safe work practices in their areas or else a particular disciplinary system will be applied.

Tell these groups the precise system that will apply to them. Be forthright and direct. Let them know that every achievement has a possible positive and a possible negative outcome, just as we rejoice in or suffer the consequences of our choices in any other effort in life. If performance is not as expected, they will be held accountable because it is *their* SH&E program. You will help them identify the areas of and the methods to achieve improvement. They are not in the boat alone. Everyone is in the same boat heading toward the same destination—excellence.

Keep everyone informed of the status of the new effort and stress the positive aspects. Especially keep everyone informed of the problems that are being encountered along the way. If the steering group doesn't share the problems, how are they going to solve them? Resistance to the solution might be just as strong as resistance to the problem. Everyone participates in this endeavor. They may not know it or accept it at first, because they don't yet understand what is expected of them. As they get more and more familiar with the new ways of doing SH&E, not only will the new ways become the routine ways, but also input from them will provide solutions to the problems. The steering committee will be surprised at the number of alternatives and solutions that will be forthcoming.

Repeat, repeat, repeat the message. Articles about the VPP effort appeared on the front page of the JSC newsletter every week; the Director of JSC issued a memorandum to all employees about JSC's plans to participate in the new initiative; and some of the members of the JSC working group set up a booth at a health fair on site to dispense VPP literature, popcorn, and balloons. T-shirts, pens, and computer wrist pads had an imprinted JSC VPP logo. Some of these activities were going on simultaneously; others were timed so that a message was continually before the employees.

The important thing is to continue the messages; regularly disseminate news about the activities, achievements, progress reports, problems, or opportunities until all employees begin to believe that this is not "just another flavor of the month" program, but is really a serious change.

Truth builds trust. Building trust comes before employees can be empowered to assume the responsibility and accountability for their actions. Truth builds trust. Honesty still remains the best policy. Describe the new system as it is, whole and complete, with no omissions or surprises to emerge when the systems are going into effect. If you don't know an answer, say so, but say you will find out what it is. Then proceed to find out the truth and get back to them as soon as you can.

If the steering committee makes a mistake, let everyone know about it and get it corrected. If what has to be said is painful, it should be expressed sympathetically and with understanding, but tell the facts as they are. By following these precepts in the communication plan, trust and credibility will be established as the new SH&E management system proceeds. This trust and credibility is required to create the employee empowerment and involvement that are essential to success.

Be alert for inconsistencies, by action or word. Initiate instant damage control if necessary, and do not try to explain it away. Tell just what happened, how it happened, and why it happened. Talk about the remedy. Then let it stand. When the employees see the remedy enacted, their attention will be diverted from the incident to the remedial actions.

Honesty and directness by the leadership and the steering committee—whether the messages are explanatory, negative or positive, or upbeat or downbeat—will convince the employees that they really are expected to participate in these new processes. Trust in the process and the leadership will be the result.

Be emotional, not just analytical and statistical. Repeated recital of the bare facts will quickly dry up employee interest. Throughout history, emotions, passions, and heated enthusiasm have roused many to rebellion, revolution, and reform. The redesigned SH&E processes are revolutionary compared to the traditional hierarchical, closed-loop departmental safety and health programs. Leadership, department representatives, and the steering committee must be evangelical and passionate about these processes that will yield such great benefits for the employees and for the company. Such passion and sincerity are contagious. Everyone needs to get excited and remain excited about the new safety and health programs. Keep the urgency heat turned to high.

Remember—walk the talk. Leadership's behavior and attitude must display enthusiasm and belief in the successful outcome of the new venture. The employees will catch the fever and march alongside. Sincerity and different voices work well, and a variety of messages from the heart will enhance communication.

If possible, send middle managers and key grassroots leaders to the VPPPA conferences along with the SH&E staff. There are annual national and regional conferences. Workshops abound on a number of topics such as benefits, participation, union involvement, and self-evaluations. By networking with other attendees, the company's attendees will experience the passion, enthusiasm, and fever that pervades these conferences. They will observe supervisors, managers, and shop operators interacting cooperatively and congenially. Leadership should also attend and hear personally how enthusiastic the participants are about how well the processes work for them.

Introduce fun and funny activities, especially if the organizational style is normally traditional and conservative. The steering committee and the communication subgroup can brainstorm ideas for games, gimmicks, and giveaways that carry a message about the processes and systems to stimulate the participation of all employees. This encourages the creative and innovative potential that will contribute greatly to actual employee involvement.

Communicate to encourage and console. These redesigned processes will cause some employees pain and trauma—fear of losing a job, fear of failure, fear of getting out of a comfortable niche, fear of the unknown. Let your employees know that you understand the trauma that they are experiencing, that they are valued and invaluable in making these changes successful, that their effort in the changes is appreciated, and that what they are doing is improving the competitive advantage of the company—which in the long term will ensure improved job security, a better quality of life for them and the company, and a safer and healthier work environment.

Communicate with specialty items or devices. Use gimmicks and touchable devices that encourage doing something new, being resourceful and creative, being actively involved before a change happens, and having fun doing it. Hammer and Stanton describe a Japanese company that used novelties made from Tyvek (can't be torn—find a way to change it), silly putty (break it—be resourceful), and so on. Accompanying each device was a message about the reengineering effort.

At JSC, the communication committee bought hundreds of little stick-on cartoon creatures (which they named Seemore Safety and Ima Hazard) that had a short piece of ribbon stuck to their heads reading "VPP and JSC." Seemore and Ima turned up stuck to something in all areas of the facility—executive offices, testing facilities, restroom mirrors, on top of computers, in machine shops, in astronaut offices, or on managers' desks—they were everywhere. The committee had small stick-on labels printed that read, "You are looking at the one responsible for your safety," and stuck them in a lower corner of all of the bathroom mirrors at JSC. There are numerous such items available from novelty companies. They will help you develop ideas about how to use their stuff to sell your new "products."

Listen, listen, listen. Lastly and most importantly, remember that communication is a two-way street. One of Stephen Covey's seven habits of highly effective people is: "Seek first to understand, then to be understood."[5] You must listen to learn how your messages are being received and perceived by the recipients. Listen to the employees' opinions about the redesign plans for the SH&E management systems. Provide opportunities for them to vent misunderstandings and frustrations, to ask questions, and to make suggestions. Not only are you hearing what they have to say, but also you are *involving them* in the process.

Listening frequently and silently also helps convince employees that this is not a flavor-of-the-month effort that is being forced upon them by management. Once they have opportunities to talk, they are then involved in the program itself. They have had their say and they feel more comfortable with the new plans. They have put something of themselves into it and now own their part of it. Continue this throughout the gap analysis and the implementation until the new criteria become the routine way of doing business.

Communication media can be town meetings, functional or departmental group meetings, and safety or staff meetings whenever and wherever they are held. At JSC, the VPP coordinator and the communication subgroup used town meetings of union representatives from the 20-plus unions represented onsite and invited as guest speakers union representatives from companies that were VPP participants. A second town meeting was held with all of the middle managers and supervisors onsite. The guest speaker was the plant manager of a Mobil Chemical plant that was a VPP participant. A third town meeting was held with the CEOs and contract program managers of the 40-plus contractors onsite. They were addressed by the director of JSC and her director of safety, reliability, and quality assurance.

A member of the steering group or the communication subgroup attended the monthly departmental or contractor safety meetings, preached the word, and listened to frustrations, opinions, suggestions, challenges, indignities, and questions. As many of these meetings as possible were videotaped or recorded on tape or by handwritten notes. The messages were reviewed and evaluated by the entire steering group to derive future actions or changes to use, appease, soothe, or answer.

As far as these actions are feasible for your organization, your leadership and your groups can follow the same path. Just remember that the talking and the listening should be done by people communicating directly with each other as often as possible. A member of the steering committee should personally attend these sessions frequently, whether it is a small meeting or a large one, and whether yours is a small, mid-size, or large company.

Surveys and questionnaires can be used to allow those who hesitate to stand up and be heard publicly to have an avenue to express themselves anonymously. Whatever you do, tailor the communication plan and activities to your organizational uniqueness.

Summary

Chapter 9 has discussed culture and communication as they will affect the stakeholders—the employees—in the reengineered SH&E processes. A case study was cited to show how a small company implemented a quality management program and that took three to five years to show results. Change of corporate culture and values requires patience and perseverance. Recommendations have been offered to ensure that your corporate culture is founded on values that will serve the company. Business values have been suggested that can be used as guidelines against which to measure your own. The chapter discusses how to get the employees to accept responsibility for their own safety. Also included are the fundamental methods and techniques for communicating the proposed changes of the culture and the SH&E management systems to the different segments and levels of your organization.

Chapter 10 addresses continuing improvement and maintaining a high level of implementation and enthusiasm.

References

1. Drucker, Peter F. *The Practice of Management.* New York: Harper & Row, 1954: 64.
2. De Pree, Max. *Leadership is an Art.* New York: Doubleday, 1989: 96.
3. Ibid: 95.
4. Hammer, Michael and Steven A. Stanton. *The Reengineering Revolution.* New York: HarperCollins Publishers, 1995: 143.
5. Covey, Stephen R. *The 7 Habits of Highly Effective People.* New York: Fireside, 1990: 235.

The Stakeholders

10 The Never-Ending Journey:
Continuing Improvement

5. Improve constantly and forever the system of production and service, to improve quality and productivity, and thus constantly decrease costs...[1] *—14 Points of Management,* W. Edwards Deming

The Dynamics of Change

The dynamics of change can be described as a belief in optimizing complete systems, in a broad, long-range focus on planning and results and in root cause analyses and system improvements. It can also be a belief in partnerships, cooperation, and teamwork, striving for win-win situations, and a focus on long-term balance with a practical recognition of near-term needs and constraints.

With the revised SH&E processes in place and implementation on a progressive schedule, leadership and staff can now concentrate on balancing the needs of the present with their vision of the future. They should also recognize that the needs of the present are in a constant state of change. This is another justification for the annual self-evaluation. Here we are specifically addressing the dynamics of the SH&E management system, but an annual review of company-wide activities will also serve well, since all operations interact and influence each other.

SH&E processes and management systems are never a static set of unchanging, written-in-concrete procedures. Some companies (for instance, petrochemical) must review their processes at least every three years under the OSHA process safety management rule. Others must review their procedures and processes every few months, such as construction companies that move from site to site. Each site will have its own conditions or operations for which safe work practices must be developed before work can start. Many of the OSHA standards require employees to be trained in job-specific safe work practices before work begins, as in confined space entry and excavation and trenching. We are surrounded by evidence of the dynamic, ever-changing environment in which we live and work.

Review and change are integral to our daily lives. We cannot ignore them if we are going to survive as a company, a community, or a family.

Confronting and solving the challenges of these changes becomes the company's continuing improvement program. These are the goals and objectives from year to year. The challenges themselves will differ from year to year; but you can rely on changes, variations, and differences occurring annually. In the SH&E world, improvements and changes are developing, being discovered, or being mandated in a constant stream. The self-evaluation, gap analysis, and implementation activities that you have conducted and implemented probably indicate that some requirements have changed, that you are not in compliance to the extent you thought you were, and that some systems currently in place are no longer effective.

By now you should recognize the need for a structured but flexible process to identify improvements or requirements not implemented, new tasks, and jobs not yet done. The process must be structured to create an orderly, systematic method to identify and solve problems and implement change. At the same time it must be flexible to meet the challenge of unknown, unexpected change. You have put that process in place through the SH&E management system that you have implemented. If properly implemented, it will ensure that nothing is omitted. It now remains to have a system that addresses the details of ongoing implementation.

Conoco, Incorporated has a Process for Continuous Safety Improvement (PCSI)[2] that also includes occupational health improvements. The PCSI focuses on four primary components: leadership, people, assets, and systems.

Under these four divisions, Conoco identified a total of fourteen core elements that provide the structure at all of their facilities worldwide for a consistent safety and health program. Within the fourteen elements they have provided flexibility so that each site can set goals and objectives, and modify, adjust, and implement changes unique to the location. Conoco's identification of the core elements is shown in Exhibit 10-1. The primary components and the fourteen core elements are similar to the VPP criteria.

Yours may not be a worldwide company like Conoco. It is probable that you are among the thousands of moderate-size companies (250 to 5,000 employees) that are not international, but may have more than one location. A company as small as 175 employees can have 18 or 20 small external service locations close to customer sites to enable it to provide fast-response service. There are consistencies and differences at every location. The SH&E management program can be managed from the corporate offices, but each external location has the liberty and authority to issue site-specific work practices to meet a particular customer's policies. For example, a customer may require that their emergency evacuation procedure be a part of the company's safety manual. The corporate office accommodates this request by appending the cus-

Exhibit 10-1. Conoco's "14 Elements of Constant Improvement"

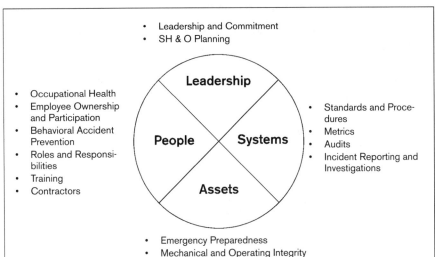

Structured to include four primary categories and major sub-elements of safety and health managment systems corporate-wide. The sites and locations have the flexibility to evaluate and improve their individual programs tailored to their work environments, but they are measured against the corporate-wide performance expectations. The expectations are in a matrix that describes five levels of achievements from 1 to 5, 1 being the lowest and 5 being the highest. The goal is to achieve 5 in all sub-elements. A similar matrix can be found in the OSHA PEP document (Appendix G).

Source: Conoco, Inc.

tomer's procedure as an attachment to the company's emergency response procedure in their SH&E manual.

Consistency is necessary to follow the fundamental precepts of the VPP SH&E management systems. You can achieve this by retaining responsibility for developing and disseminating SH&E management policies and procedures. Implementation of the policies and procedures, however, can be the responsibility of the management and employees of each location, department, or region, so long as they remain within the parameters of the corporate requirements.

The Evolving Improvement Process

During the first one or two years, leadership and the steering committee have built the plan of action, and have learned how to implement it and how to work together as a team. After the initial learning phase, the next steps will be easier and more efficient. Now the improvement activities of the safety and health management processes begin.

After the annual assessment and evaluation each year, you will have a new set of accomplishments to recognize or deficiencies to fix. During the annual evaluation, learn what progress has been made with the new systems and processes instituted the preceding year; what changes or adjustments should be made to make them better; what new systems or processes developed during the past year—whether OSHA or company-based—must be added to the SH&E programs; and what incomplete tasks should be addressed.

Not only that, you can reflect upon and give more consideration to issues that were left pending during the initial planning and learning phases. These are the items that were left over when leadership and the steering committee assigned priorities to the gaps requiring immediate closure. At this stage, during the planned annual evaluation phase, you have the luxury of examining only the changes and improvements that are indicated.

You do not have to worry that the processes' structure may be incomplete, or that some process or system is missing, because the VPP criteria require comprehensive programs in the four critical elements. The programs are generically structured so that they cover any operation or function in the company. Be sure to emphasize this condition to the steering committee—no revision of the processes or systems is needed unless the annual review indicates they are not meeting expectations, and indicates where they should be improved or updated. This holds true, of course, only if major course corrections in the basic processes and systems have not been mandated by OSHA or by a powerful external market force. If that has happened, then adjustments and modifications will have to be initiated to meet the new course corrections.

Use the VPP Evaluation Guidelines (Appendix D) as the benchmarks to measure your progress. The guidelines assure credible and verifiable findings about the systems on which you can base your goals and objectives for the coming year. Here again, for the new goals and objectives, leadership and the steering committee go through the process of determining hazard potential, realistic costs, and projections to completion; and setting priorities and schedules. Be sure to schedule the analysis of the annual evaluation findings to coincide with the budget cycle so that funding for the new goals and objectives is folded into the overall company budget for the coming year.

The process of plan-do-check-act becomes the norm each year. Once it is repeated through several cycles it will become routine. When it becomes the norm to plan for it, schedule it, do it, and act upon the findings on a yearly basis, you have accomplished values and attitude changes that were planned at the start of modifying the SH&E program.

Exhibit 10-2. Completed Staff Work as Presented by Stephen Covey

1. Provide a clear understanding of the results.
2. Provide a clear explanation of what level of authority you are granting.
3. Clarify assumptions.
4. Provide as much time, resources, and access as possible.
5. Set a time and place for presentation and review of the completed staff work.

To get timely responses, set deadlines for the completion of proactive items. To ensure that deadlines are met, use the practice that has been taught in business administration courses for years—the "completed staff work" method to set goals and get results. The five steps identified by Stephen Covey in his book, *Principle-Centered Leadership*,[3] are shown in Exhibit 10-2.

In the paragraphs that follow, we have applied these steps to the gap closure and implementation tasks relating to the continuing improvement of the SH&E program management systems.

With regard to providing a clear understanding of the results when the strategic plan was developed and the gap analysis was compared to the current safety and health program, leadership and the steering committee defined the results that were expected. Each program element subcommittee knew what their gaps were, what the strategies were to close them, what options or alternatives might be available, and what the outcome should be. At the outset of the closure activities, be sure that assignees for gap closure understand these points.

Ensure that each subcommittee and individual member of the working groups know what level of authority that they have to close the gap or correct the condition. There will be employees of different functional levels on the committees. There may be plant superintendents, operators, administrative assistants, departmental managers, union shop stewards, and others. Each one functions at a different level of authority. This is a critical concept of employee involvement to impart to all employees. Each one at his or her level exercises his or her level of authority to close a gap, to correct an unsafe practice or condition, to conduct an inspection, to investigate an accident, or whatever the task may be.

The shop operator can take an unsafe ladder out of service and report it to the foreman. The foreman of the shop can authorize purchase of a new ladder. The plant superintendent can initiate a plant-wide inspection of all ladders. The vice president of operations can recommend to the chief executive replacement of all defective ladders and a continuous replacement program based on regular inspections. Any one of them, from the shop operator to the vice president of operations, can recommend regular training in how to inspect and use ladders. These are the different levels of authority that should be granted to your employees. The important aspect is that leadership imparts to them the knowledge and assurance that they *have* the authority to exercise. This is how a system for a safety program is built. It has different levels of actions and applies at all levels.

When the gap closure and implementation activities begin, ensure that the employees are working with the same set of assumptions that leadership has. They should know what the expectations are, what the OSHA requirements are, what will meet the OSHA requirements, and what outcome signifies success. Communicate this information from time to time through newsletters, training sessions, safety meetings, and other media. Encourage questions and requests for clarification at any time.

With regard to time, resources, and access, be sure that the committee members and the employees understand when a gap or implementing action *must* be completed, what resources they have to complete the task and how to obtain them, and who or what is accessible to assist them in getting the work done.

It is also important that they have all of the relevant information about the task or project. For example, the training subcommittee starts developing a plant-wide training effort in hazard recognition and control tailored to the hazards by department or operation. They identify instructors, develop lesson plans, reserve classrooms, and are about ready to set up schedules. Then they put in their requisition for audiovisual equipment and lesson materials, only to be informed that the money is not available at this time. This should have been known up front. So, be sure that all who should be involved *are* involved.

Before the project, task, or gap closure is put into effect, schedule a specific time, date, and location for the committee or the employee to present the completed assignment to the appropriate leadership. Have periodic reviews if the task is company-wide, critical, or sensitive, to ensure that it is tracking in the desired direction. The committee members or employees working on these projects want to know that they are on the right track. Give them an opportunity to learn where they are and what leadership thinks about it. Put the review on your calendar for a specific date and time.

If this method is used for the daily tasks related to long-term goals, you will make progress one step at a time—another way to achieve short-term gains.

Building and Maintaining Employee Interest

You will be constantly challenged to build and maintain employee interest. Unless you beat the drums of communication and recognition regularly, interest will soon wane. Consider how many whims and novelties pass through our lives from year to year—hula-hoops and trampolines, for example—that are soon to be a memory. Interest quickly fades over faddish ideas.

One of the most important actions that leadership can continue is to recognize and reward new efforts, creative ideas, and innovations contributed by employees during the year. Not only leadership, but also employees, are the leaders and winners in applying and practicing the VPP criteria. Once they experience the exhilaration of making decisions, implementing them, and achieving the desired outcome, they will be the staunchest supporters, the most creative innovators, and the best performers for continuing involvement in the SH&E management systems and processes.

There are so many tasks related to making these systems successful that you need all the willing helpers that can be identified—not only to make the processes function successfully, but *to get the work done*. The more people who share the multitude of tasks, the lighter will be everyone's load. During the early years of implementation, they will be performing three essential tasks almost simultaneously—today's routine tasks, the SH&E-assigned tasks, and tasks related to the strategic planning process for the next two-, three-, or five-year cycle.

A Mobil chemical plant manager stood before a VPPPA conference and described how VPP participation was estimated to have saved Mobil $10 million since being accepted into the program. When the author interviewed him later, he was asked how he got all of the work done relating to VPP and still managed to do his daily work of keeping the plant operating profitably.

His reply was:

> I didn't do it by myself. I didn't even try to do it by myself. In the first place, it would be an impossible job. In the second, that isn't the way it is supposed to be done at all. Involvement is the key element. I started every morning with staff meetings of my managers. They were assigned certain jobs to pass on to employees. We spread the tasks as much as possible. Besides, I decided early that if I was going to have to work that hard, so was everybody else!

The Never-Ending Journey

This brought a laugh from everyone, but it's true. The details are numerous, and you need everyone working just as hard as leadership. It doesn't matter whether it is a small business location or a large one, the tasks are the same—it is only the application of them that differs.

Examples of sharing the workload are published in the "Best Work Practices" feature of the VPPPA publication, *The Leader*. The features for the Summer 1997 and Winter 1998 issues are Exhibits 10-3 and 10-4. The two questions asked in these issues were:

- "How did your site implement an effective safety committee?" (Summer, 1997)
- "How did your site involve employees in conducting the annual self-evaluation?" (Winter, 1998)

Exhibit 10-3. Best Practices Exchange, *The Leader*, Summer 1997

Question: How did your site implement an effective safety committee?

Number 1

Answer: We had a good safety committee prior to applying for the Voluntary Protection Programs. All employees and management participate on the committee and are very cognizant of safety issues. The VPP program gave the committee greater authority. The committee now participates in our work order system, which allows employees to report safety and health concerns and ensures employees receive feedback on what actions were taken to address their concerns.

Jean Thomason
International Paper Container Plant
Bay Minette, AL

Number 2

Answer: Aker Gulf Marine's Safety Committee has been very active in a vital part of our successes in the VPP Program and our overall safety efforts, including our Committee Chairman, Jerry Andrews of the Painting Department. The members represent all the different crafts, a supervisor, an engineer, and management.

The committee meets weekly, and each meeting rotates from a conference room meeting to field inspections the following week. The members also join our Safety Superintendent of the Week as they inspect different areas each week.

Each member serves one year and his/her terms are staggered so that two new members are added each quarter. The committee selects its new members from a group of volunteers from each department. This keeps "new blood" on the committee as well as maintaining a consistent flow to the group.

Don Goodwin, Safety Supervisor
Aker Gulf Marine
Ingleside, TX

You will note in the responses that participation by the employees came from all levels and responsibilities. There are many ways and opportunities to involve employees in the daily activities of the SH&E systems. Ask them; they will tell you. Some of them have favorite complaints that they want to get off their chests—put them to work finding a solution. Some of

Exhibit 10-4. Best Practices Exchange, *The Leader*, Winter 1998

Question: How did your site involve employees in conducting the annual self-evaluation?

Number 1

Answer: Gardner Cryogenics' Joint Union/Management Safety Committee directs the facility's annual self-evaluation. The committee consists of four elected union representatives and four salaried representatives. Our plant has about 150 employees. Union safety committee members select six hourly employees, representative of the different job classifications from each shift at the main plant, and six from our second, smaller plant. One of the salaried committee members selects six people from the salaried staff to participate. In all, about 30 people are asked to rate Gardner's safety performance in each of the 17 self-evaluation categories. The safety team members guide the participants through the evaluation process by answering questions and explaining what we are doing in each of the self-evaluation categories. We buy lunch for the participants and do this in an informal, relaxed atmosphere. We also ask the participants to comment on what they feel Gardner does well in safety and what safety concerns they have. Employee responses to the self-evaluation are confidential. All information is tabulated and each of the categories is scored based on the participants' responses. We then use the self-evaluation, along with injury report and near-miss data, to develop our safety improvement plan for the year.

Wayne Petel, Safety Manager
Gardner Cryogenics, Bethlehem, PA

Number 2

Answer: In the past, Mead's Annual Program Evaluation (APE) was conducted by the Division Safety and Health Manager and me. In 1997, it was decided that each plant in the Containerboard Division would be responsible for conducting an internal self-evaluation. In selecting representatives, I was looking for two employees who had previous involvement in safety and health programs, and auditing experience. After talking to a few candidates, our team was formed. The three of us divided up nine major categories for the evaluation: (1) Management commitment and leadership; (2) roles, responsibilities, and accountabilities; (3) employee participation; (4) communication; (5) behavior skills; (6) hazard recognition and control; (7) education and training; (8) incident investigation and analysis; and (9) performance measurement. As a result of the evaluation, some excellent recommendations were made and corrective actions implemented.

Rick Wrobel, Quality Assurance Manager
Mead Containerboard, Bridgeview, IL them have ideas that they would like to try on the job to improve production—tell them to prepare a plan for it (how it is to be done, what the estimated cost is, what the results should be) and come to see you.

An improvement in the smallest task is an achievement. It is an advance that you did not have yesterday. It is also employee involvement, even though it may concern the smallest tasks. Be positive if their ideas do not work. "You tried. It didn't work. Don't stop. The next one will work, or the one after that, or the one after that. Don't hesitate to bring me other ideas." Be supportive of those who fail. Encourage them to try again and again. Recognition and a pat on the back is invaluable and greatly motivating.

The Geon Company, LaPorte Intermediates Plant, LaPorte, Texas, was approved as a Star site in September 1997. In a letter to the VPPPA, Bill Lesko, then manager of safety and health at the plant, commented that employee involvement was one of the obstacles during their application process. As the effort got underway, however, the employees began to show more interest and volunteer more help. Lesko said, "People who never said more than the minimum were now making suggestions and volunteering to help…Safety is no longer solely owned by the safety manager or the plant manager. The LaPorte Intermediates Plant now has 180 safety managers working for the Geon Company."

Ensure that new employees get inoculated with the corporate values and objectives. Safety, health, and environment come first, period. Sometimes your efforts fail because of failing to pass on the company philosophy to new employees. If you do not now have one, begin a new employee orientation program that not only is orientation in job skills, but also includes company values, beliefs, goals, and objectives. The orientation should cover both the SH&E management objectives and the objectives of the company to improve the quality of life for all.

Rotate the steering committee members

Rotate the steering committee members periodically. The original members can get tired or lose interest. They have become veterans now in implementing the plan and getting it into action. They might leave the group because of reassignment or retirement. The ones who are reassigned or who return to their former duties can spread the good word about the systems and processes.

Some of the original members with special expertise, who were brought in on a temporary basis in the early stages, can now be replaced with others like them or others with different skills and abilities more suitable to implementing the maturing processes. Select capable new members who are respected by the existing members. This will assure only a minimum amount of orientation in team-building skills. New members need to work alongside the veterans for a period to learn the details of the processes and systems and how they work. Retain enough veterans so that continuity of purpose and goal is maintained, but give the newcomers the opportunity to participate in the planning, decision-making, and implementation. This may be a heady experience at first for the new members, but also motivational in raising their interest to a high pitch.

The core group of senior leadership should remain constant unless transfer or reassignment occurs. The core group is the backbone of the plan—the core group being top management and the senior SH&E, product, resources, information systems, finance, and plant maintenance staff, and any other person unique to the company or industry.

The core group is the fundamental, solid authority that firmly supports the details of the systems and processes. Leadership provides the guidance and counseling to keep the project on course. The rotation procedure lets all of the employees know that the planning activities are not being reserved for just one elite group, that all of the employees over time will be given an opportunity to participate, that it is not a secret society, and that the SH&E management system is here to stay.

To ease the pang for the former members leaving the committee, have a recognition ceremony to give them credit for their pioneering work, and reserve the right to call them back for special task forces to help with the many projects associated with implementation and continuing improvement.

Encourage teamwork

During the gap analysis and evaluation phases of the redesigned SH&E management systems, establish certain principles and protocols for effective teamwork among the committees and subcommittees. When leadership adopted the VPP criteria—especially employee involvement, ownership, responsibility, and accountability—they also established principles that are basic to productive teamwork. Selecting qualified committee members who are respected and liked by other members greatly helps to build an effective team. Members who can exercise mature conciliation skills and work tactfully through differences and conflicts to reach common agreement are invaluable to the success of the effort.

Establish trust and trustworthiness in and among all of the team members, and ensure that all team members understand and accept, with good will, the decisions that leadership or the committee chair must make if consensus cannot be reached.

Celebration and hoopla; recognition and reward

We have mentioned celebrating, recognizing, and rewarding several times. One important aspect of achieving success in a venture that requires a tremendous exertion is the recognition for getting it done. Some Olympic athletes may be engaged in their sport for the fun of it, but most of them are striving to be the best. The best receive the accolades, as they should for the hours and hours of effort they have expended. Don't forget, "Everyone wants to be great." You don't know you are great unless people you respect tell you that you are.

In *The Culture of Success*,[4] Zimmerman and Tregoe discuss what strong motivational forces financial awards and individual recognition are in reinforcing company values. A number of companies were asked, "How are behavior and actions that support basic beliefs recognized and rewarded? Circle those that apply." Nine responses were listed with the question. The responses listed different cash awards, various individual recognition elements, and no recognition. The two responses ranking the highest were, "Special noncash award" (21.5%) and "Individual recognition by supervisor" (19.5%). Next highest was, "Recognition in company publications" (13.0%). Responses involving money awards ranged from 12.3% to 4.8%. Companies are becoming aware of how well positive reinforcement sustains corporate values and beliefs. And, surprisingly, it does not have to be linked to monetary rewards.

There should also be recognition and reward for innovative ideas and results above and beyond expectations. Stephen Brown's comments about incentive programs (see Chapter 9) are examples of recognition for individual efforts. There is also recognition of everyone. At the end of the first year, the measurement indicators may not show much progress. But in the second, third, and ensuing years, beneficial results will be noticeable. This is due to safer and healthier performance by all. Since everyone is involved, all should be recognized in some way. There can be such things as picnics, SH&E fairs (balloons, popcorn, and cotton candy!) to create a festive atmosphere, along with the educational activities. It can be as simple as a backyard barbecue or (if your budget permits) as imaginative as a company-paid three-day holiday aboard a cruise ship. Whatever it is, make it fun, different, and educational.

Since 1995, Johnson Space Center has had a Safety and Total Health Day devoted to safety and health activities. It is a total stand-down from the routine tasks of the day. Part of the day is devoted to departmental needs and interests planned and presented by management-employee departmental teams. The rest of the day is devoted to exploring the personal safety and health questions and concerns that the employees may have. See Exhibit 10-5 for an example of JSC's program for a safety stand-down day.

A variety of organizations, associations, and community groups that will participate in activities such as this are available in your community. Most of

Exhibit 10-5. Johnson Space Center's "Safety and Total Health Day"

For the Safety and Total Health Day on October 15, 1997, the Space Center involved all of the managers and employees, including contractors, at the site, and many community and national groups.

The Safety and Total Health Day Committee divided the planned activities into eight categories:

- Safety programs and services
- Total health programs and services
- Safety programs for specific operations
- Safety outside the workplace
- Healthy living at home and work
- Emergency planning and responses
- Vehicle and traffic safety on and off site
- Special safety and total health day activities

Four to five months before the date, the planning committee began their work. The committee had seven subcommittees composed of a total of ninety employees to plan and coordinate the activities. They enlisted the support of associations, private companies, community groups, federal and municipal offices, and emergency response groups. Sixty-four booths were set up on the JSC grounds. Represented were:

- Community and county partnership groups such as job mentoring for at-risk children, organ donation, blood drive, Houston's Drug Free Business Initiative, Auto Crime Task Force, Big Brothers/Big Sisters, outreach organizations helping the needy, child safety, driving safety, and boat safety.
- Emergency preparedness and response groups: the local Emergency Planning Committee, hurricane preparedness, Greater Houston 911, spill response team, JSC Emergency Operations Center, Houston Hazardous Materials Response Team, and the Hermann Life Flight helicopter.
- Health and fitness programs: massage therapy, cancer awareness, JSC Total Health Employee Wellness Program, Harris County Health Department Services, and the Nutrition Intervention Program.
- Organizations and associations: Alzheimer's Association, United Cerebral Palsy, American Red Cross, Texas Society to Prevent Blindness, and the American Society of Safety Engineers.
- Community law enforcement organizations: Houston Police Department, Department of Public Safety, Harris County Constable Precinct 8, Pasadena Police Department Community Service, and Houston Fire Department.

Seminars and presentations were also conducted on women's self defense, gang intervention, managing workplace pressure, CPR training, driving and drinking consequences, and violence in the workplace. A puppet show for the children was included. A fun run/walk was held in the afternoon. A concert was presented during the lunch hour on the lawn area outside one of the cafeterias. Safety and health videotapes were presented all day on the JSC closed-circuit TV channel.

them are national groups or similar community organizations. Your day of celebration can be as extensive as the Space Center's was, or less extensive, to fit your business and what you can afford. The extent of the Space Center's activities indicates just how large a number of public and private organizations are available to any employer who wants to include them in company activities. This is also an excellent avenue to display dedicated corporate citizenship and establish cooperative relationships with the community and municipal organizations.

A safety and health day is certainly a great way to get employees *involved* in this new SH&E management system. It is also convincing to anyone who may have doubted that the program is *here to stay*. Many VPP participants are having these annual stand-down days, community-company days, or other special days just for SH&E awareness, on and off the job.

Remember to celebrate the small gains—no matter how small

Plan your small gains. Don't expect them to happen by chance—chance may not be on your side. These serve as a means to keep the project alive to the employees. Even though small and short-lived, a short-term gain has notable features. First, it is familiar and visible to all of the employees. They know in an instant whether the gain is *real* or just management hype. Second, it is clearly accurate and true; there can be no argument about its status. Third, it is directly connected to the change effort going on within the company. It cannot be related to any other management project that may be in action at the same time.

Select a small project, a small improvement instituted company-wide, or one of your indicators such as a recordable incidence rate, or a lost workday rate, or noteworthy responses to your close call program—anything that will show a tangible, real change in the safety and health program performance. Publicize it, post it on all of the bulletin boards, have free coffee and cookies all day at the cafeteria or coffee wagons around the plant. Call attention to it as a milestone worthy of recognition.

Not only do these small gains have certain characteristics, they serve several important roles. John Kotter in his book, *Leading Change*,[5] devotes a chapter and several references throughout the book to the importance of small gains. He identifies six positive effects that can accrue through recognizing and publicizing small gains as they occur. He suggests that you look for possibilities in company activities to realize these small achievements. This is easy in the company activities discussed in this book. Opportunities for small achievements are plentiful in gap closure and the implementation activities of subcommittees.

Small gains and improvements encourage the implementers to try other ideas and possibilities. Successful change prompts more change that, over time,

Exhibit 10-6. The Role of Short-Term Wins

Provides evidence that sacrifices are worth it. Wins greatly help justify the short-term costs involved.

Reward change agents with a pat on the back. After a lot of hard work, positive feedback builds morale and motivation.

Help fine-tune vision and strategies. Short-term wins give the guiding coalition concrete data on the viability of their ideas.

Undermine cynics and self-serving resisters. Clear improvements in performance make it difficult for people to block needed change.

Keep bosses on board. Provides those higher in the hierarchy with evidence that the transformation is on track.

Build momentum. Turns neutrals into supporters, reluctant supporters into active helpers, etc.

leads to the total vision being realized. Successful grounding of change in the company culture comes about by a continuous sequence of one small gain after another.

Sustain the momentum and urgency

Do not relax, however. Momentum and urgency must be maintained to sustain and nurture the new culture. The roots are young and tender. Resistance is waiting like a weed to spring up and overpower the young growth. The authors whose publications have been consulted while writing this book (see the footnotes and the bibliography) cite example after example of the failure of culture change because of letting go or letting down too soon and assuming that predetermined behavior is equal to permanent change.

Not so. The authors agree almost unanimously that grounding cultural change firmly in the foundation of a company's operations is difficult, tedious, and demanding. It takes years to complete the transformation. Persistence and patience must be exercised. A leader, a corporate conscience, or someone who thinks in the long term and is willing to make the commitment to stay with the plan, is the one required to direct this venture to completion. Managers are prone to think in the short term because they are concentrating on the daily tasks to achieve immediate gains. This is what they are supposed to do. Leaders plan the future and can wait to achieve the long-term goals—in five, ten, or even twenty years. In some cultures of the world, a hundred years is not considered too long to wait for a plan to materialize.

One of the famous Rothchild family was an avid horticulturist and gathered flora and fauna from all over the world. His gardens and collections were famous. He set out to find a particular tree that was hard to grow and slow to reach maturity. When he finally found one, he rushed into the garden house and told his gardeners, "Come, come, we must plant that tree today."

"But sir," the gardener replied, "it will take that tree two hundred years to reach full growth."

Mr. Rothchild replied, "Yes, yes, I know. We haven't a moment to lose. We must start now."

Mr. Rothchild not only had the short-term eagerness and the long-term patience, but he had the commitment to take the time to develop healthy and deep roots. He knew he wouldn't see the tree full-grown, but he wanted the pleasure of watching it during his lifetime, and he wanted it to be as big as possible before his life was over.

Keep the pitch of urgency high and heated. Every year governmental agencies or economic conditions issue new or revised requirements for the business community. Some of these are beneficial; some are burdensome and costly. It is well to stay alert to these encroachments and to evaluate their effects upon the company's economic health. Monitor your trigger points regularly. Proaction is more efficient and productive than reaction when you can foresee what the impact may be. Remain relentless in urging that today's proactive tasks be done expeditiously and remain vigilant in watching for the future tasks that may require leadership's visionary capabilities. Share this visionary thinking with the employees to keep them informed, interested, and involved in making these new programs, as well as the future ones, work to their and the company's benefit. Plant your trees today.

Keep improving—stay up with the times

Do not allow plans or systems to grow stale. Stay current with conditions in your industry and market, in the economy, and in federal activities. The annual self-evaluation indicates how well processes are working and what revisions are indicated. Leadership has identified the trigger points that point to fluctuations in the business world and in the federal or state SH&E activities. Stay up with information systems and technological advances that can improve the SH&E management processes, or any other processes for that matter. Continue to benchmark other companies or competitors in the industry. Stay abreast of the ISO requirements. They reflect alterations in present or projected future international conditions, which can affect the interpretation of their requirements. This can, in turn, have an effect on how your systems and processes comply with the altered criteria.

Maintain leadership credibility—continue to walk the talk. Continue to set the example for safe and healthful conduct. Steadfastly continue to apply the values to decisions. Take frequent opportunities to sharpen managerial skills. Develop a positive and enthusiastic attitude, improve communication skills, learn how to "make friends and influence people," how to resolve conflict, and how to develop effective and productive teams.

Do not fail, personally, to put into practice the company values and beliefs. They frequently become old hat over time. Leadership may think that everyone knows by now that they fully support the values, and that it is no longer necessary to be so visible. Not true. After a few months or a year, if leadership begins to ignore or slack off on the behavior that reflects the company values and beliefs, such as forgetting their hard hats when entering an operating unit or not even going into operating areas at all, leaders will be surprised how fast employees follow their lead. No matter how much leaders may think that they are not being observed by employees, or that employees are not paying any attention to what leaders do or fail to do, *leaders are being observed every day because they* are *the leaders.* Leaders set the pace of the march toward the company's future, and employees follow. Do not underestimate the influence that management has over the attitudes and behavior of the people who work for the company. As the true corporate consciences of the company, they assume one of the most important responsibilities of leadership: always set an example of how they want the employees to believe and to behave.

Communicate, communicate, communicate

Continue the communications—people have to hear the same message over and over before it penetrates the recesses of their consciousness. Refer again to the methods described in Chapter 9, "Stakeholders: Culture and Communications." Stay personally involved in the communications program. Ensure that managers and supervisors are the ones to communicate the corporate beliefs and values. Zimmerman and Tregoe, in *The Culture of Success,* recount the example of Bernard Marcus and Arthur Blank, the founders of Home Depot, who continued to lead training sessions even though at that time the company had 70,000 employees. Marcus is quoted, "Nobody else does training this way. It's time consuming, it's hard work." The authors continue, "The CEO is chief training officer? Get used to it. How else do you instill the right culture in a company?"[6]

Use the communications media and avenues that have been developed. Keep the messages coming regularly and frequently. The day-to-day mentoring, counseling, and conferring that occur between managers and employees provide opportune occasions to convey beliefs and for management to set an example. One author calls this imparting "positive energy," not just skills and networking.

A steady stream of communications about beliefs and values reinforces them in the minds of the employees. Observing that their supervisors, managers, and CEO are not only talking beliefs and values, but are practicing them, influences the employees to follow suit. When the employees practice and talk about beliefs and values, their behavior influences other employees, who influence others, until finally the beliefs and values extend throughout the company. Continuous reinforcement is the key element to make this happen.

Do not ignore or forget OSHA's rules. They are legal obligations and they must be attended to on a daily basis. When compliance with the requirements becomes second nature and employees feel it is the only right way to work, the company has achieved a landmark to be hailed. To get to that point, keep the employees totally involved in deciding and administering the daily, routine, and non-routine jobs and projects. So long as they have ownership, they will respond positively and productively.

Summary

The dynamics of change has been discussed in this chapter. Corporate leadership can expect the challenges of changing conditions in industry, in the economy, and in governmental regulatory activities at both the state and federal levels. Review and change are with us constantly. We must expect and learn to cope proactively with these dynamics. Conoco's Process for Continuous Safety Improvement in SH&E matters was mentioned and we pointed out that their fourteen points of evaluation and analysis closely resemble the VPP criteria for the safety and health management systems and processes. It is suggested that, to maintain consistency in programs, management retain management of the corporate SH&E systems, but allow supervisors and employees leniency in departmental or operational application.

The evolving continuous improvement program from year to year was identified because of the dynamics of the work environment; but the plan-do-check-act process provides stability to evaluation and resolution of the dynamics. Methods were suggested to maintain the momentum, urgency, interest, and enthusiasm of management and employees. Recognition and celebration activities were described to emphasize the importance of safety and health, both in the workplace and away from work. Rotation of the steering committee is suggested to provide employees at all levels with participation opportunities and to assure the employees that it is not a secret society of an elite management group.

Planning and using small gains to create and maintain interest is recommended, as is constant and frequent communication about the programs and systems. Last, but not least, leadership must meet the challenge of keeping the program alive and staying attuned to the volatile conditions in the economy, in industry, and in governmental affairs. Being aware of the conditions that create change can make the continuing improvement process one of prepared and proactive anticipation and unknown, but certain, productivity.

References

1. Deming, W. Edwards. *Out of the Crisis.* Cambridge, MA: MIT Press, 1986: 49.
2. Maniatis, Dean. "Safety and Occupational Health Management Systems," *Proceedings*, 1995 Safety & Health Occupational Health Forum. Houston, TX: Conoco Incorporated.
3. Covey, Stephen R. *Principle-Centered Leadership.* New York: Summit Books, 1991: 240–41.
4. Zimmerman, Sr., John and Benjamin B. Tregoe. *The Culture of Success: Building a Sustained Competitive Advantage by Living Your Corporate Beliefs.* New York: McGraw-Hill, 1997: 80–81.
5. Kotter, John B. *Leading Change.* Boston: Harvard Business School Press, 1996: 117–30.
6. Zimmerman and Tregoe. *The Culture of Success.* 58–59.

The Never-Ending Journey

11 Epilogue
Organization of the Future

Nothing fails like success. We can abridge all of history into a simple formula: challenge/response. The successful response works to the challenges. As soon as the stream changes, the challenge, the one successful response no longer works. As new challenges arise...the response stays the same because people don't want to leave their comfort zone...there is simply no short cut to developing the capability to handle with excellence almost any situation or condition that might occur...Real excellence does not come cheaply. A certain price must be paid in terms of practice, patience, and persistence—natural ability notwithstanding...

—Stephen R. Covey, *Principle-Centered Leadership*[1]

The model safety and health program, to which we have applied traditional and new organizational philosophies and techniques, is not a localized change in how to manage your SH&E systems and processes. It is not confined to one operating location or one department, nor is any office or business area exempt. The model applies company-wide. It affects all of the company's people skills, process operations, workplace environments, and basic management premises. At this stage of strategic planning, the new SH&E management system should be evolving into routine systems.

You probably have already noticed different SH&E activities occurring in offices, warehouses, machine shops, and computer rooms, and noticed that they are no longer considered different. Maybe there have been unexpected and momentarily unpleasant bumps along the way. These are to be expected with any activity outside of the familiar routine. People are not yet trained, or they are trained but their skills aren't sharpened, or the system needs to be modified, but they are making the effort. The transformation, however slow, should be rewarding to observe in action.

And you have the assurance that the transformation emanates from universal principles that have survived over time. Your reengineered SH&E management system is grounded on consistency of content and constancy of purpose. The results of the systems and processes required by the VPP are, predictably, intended to be excellent. With confidence, you can depend on the system, when fully implemented, to yield verifiable results of excellent performance.

During this period, corporate leaders may have had to stretch themselves to envision and lead the effort required to implement the VPP criteria as the company's safety and health management system. One of the primary criteria of the VPP is continual improvement. By definition, this means to employ innovative and creative implementation, which in turn yields beneficial and effective results.

OSHA expects programs and systems beyond mere compliance. The criteria intentionally stretch you to be innovative, creative, and flexible to attain the absolute best performance from your people in the safest, healthiest, most motivating work environment that you can possibly realize. You are challenged to look at your company's world differently. This is not the only difference to expect in the future.

Conditions of the Future

Futurists are unanimous in foreseeing an entirely different world than the one in which we are now living. Today, many conditions exist in our business and personal lives that did not exist even ten years ago, much less twenty or thirty. There is the explosion of information technology that provides new ways to manage change. What this expansion of information flow has done is broaden our worlds from our neighborhood or town to countries on the other side of the globe.

Let's suppose that you have a small machine tool business that produces precision-made instruments for use by electronic technicians, dentists, physicians, or optometrists. Your most aggressive competitor may be in Switzerland, Japan, or Germany. This competitor probably has a communication link (whether or not they realize it!) with every one of your customers here in the United States. Not only that, transportation and delivery of worldwide shipments is being expedited by the new technology. Your global competitor's delivery of an order may be just as timely as yours. Even with long-distance delivery, his prices may be lower than yours, possibly because he does not have the strict federal compliance rules that add to the cost of doing business.

How will all of this affect sales and internal programs that must be controlled to maintain a reasonable profit? With strategic and proactive planning, you are well prepared to respond to these global challenges with strategies and tactics that will help your business grow and prosper.

The first two chapters of this book discuss the costs of inadequate control and management of safety and health programs—increasing workers' compensation costs, high turnover, product waste, and intolerable production. Perhaps you thought when you read them, "Are you trying to tell me that if I change the way my safety and health programs are implemented and managed, that I am going to start making a profit every year, without fail?" Of course not. Implementing the VPP criteria as the SH&E program will not cause you to turn a profit if the other operations of your business are not functioning as they should.

Pat Horn commented when we were developing the outline for our first book, "VPP will not save a company that is going downhill. The criteria are intended to be implemented by employers who have the foresight to understand that their safety and health programs must be as productive and efficient as *every* other function of their business. But too many employers overlook the negative effect that a less-than-adequate safety and health management system can have on their profit numbers."

We have pointed out the positive financial impact that an excellent safety and health program can *contribute* to the optimum performance of your company as a whole. The VPP criteria, as written by Peggy Richardson and Pat Horn, intentionally permeate every operation and function in a business enterprise, because that is the way that an excellent SH&E program *should* work.

There is no indication that the costs discussed in the first two chapters will decrease any time soon. There is every indication that they will *increase*. Health insurance and medical expenses are growing at astronomical rates. Yet, as has been pointed out, the work environment is getting safer each year. Every proactive, preventive procedure, method, or technique that you introduce into the SH&E system is a means of controlling these unwelcome costs.

Organization of the Future

To respond to the accelerated pace of today's and tomorrow's business world, companies must be quick, flexible, and adaptable—in other words, prepared for whatever comes. The world as we know it is changing at an extremely rapid rate. All of leadership's capabilities as leaders will be tested to the maximum in future years.

There are other features of tomorrow's companies about which futurists agree. Tomorrow's companies will be service-oriented and responsive to their customers and stakeholders. More will be people-oriented and less functionally organized. Employees will be empowered to assume responsibility and accountability for their piece of the company's operations. Employees will not fit into boxes (organization charts) anymore. They will be as informed as management about the company's operations, goals, objectives, missions, and cash flow. They will cooperate with the leaders to work smarter and more productively.

Employers and leaders are placing more responsibility and accountability on the employees who make first personal contact with the customers. The employees are being trained to ensure customer satisfaction. They will be empowered, self-directed, and informed. They will know what the company's goals and objectives are, how to contribute to them, and what it takes to achieve them. Employers will seek opportunities and methods to provide the employees with more professional and technical career development, will be dedicated to establishing a flexible and supportive work environment, and will allow and encourage employee participation in work design and implementation. By giving employees more, you will get more—trust, loyalty, *and* productivity.

The truly successful organizations will be circular in construct, more team-oriented, and more open to candid information exchange. This arrangement facilitates problem solving, optimizes opportunities, and resolves challenges. We must have organization to get our work done, but it does not have to be structured as it is now, nor will it be in future years. Remove the boxes and draw arrows to *and from* the functions. There will be no more tiered, box-like organizations; instead, collaborative, multifunctional, discerning, and knowledgeable teams will get the work done faster, more efficiently, and more productively.

The companies that will flourish in this nimble, quick, and flexible environment will be focused on and will exercise certain specific qualities. All members of the company will share and be committed to the mission and vision for the company. They will be committed to doing the right thing effectively and productively. The leadership will promote, encourage, and coach all members to seek and to learn new skills and new technologies, and to apply their new learning creatively to their work lives as well as their personal lives. Along with conducting an annual self-evaluation of how your SH&E management system is performing, in future years you will also include an assessment of the organization's culture and change it as needed to ensure that it maintains viability, vigor, and flexibility.

The Economy of the Future

The economy itself will change—it is changing as we write—to one that is technologically based, service-oriented, and reaching more markets worldwide. In some sectors of the economy, new jobs are being created past the capacity to fill them (the information systems industry, for example) for programmers, technicians, and better educated, highly skilled employees.

Everyone in the company must be computer literate, including leaders and managers. Even if the company is medium size, you are in the era of computerized business functions. Leaders and employees are constantly acquiring new equipment, new electronic capabilities and software, new knowledge, and gaining new skills to enhance the quality of service to your customers. Be aware that these new skills and equipment also have their attendant hazards. To cite a few: significant hazards have surfaced (and will continue to do so) in ergonomics, computer work stations, carpal tunnel syndrome, and exposure to a multitude of toxic chemicals, not only in the chemical production units, but in the use of the products. Fires in homes and public facilities, such as hotels and restaurants, can release extremely hazardous materials.

During the self-evaluation cycle, pay special attention to the emerging processes that are capable of producing new gadgets or facilitating existing processes. Note the resulting changes in the workplace and evaluate the related hazards—*prior to their use or introduction into the workplace*. Start control measures immediately and proactively.

The Future of SH&E Compliance

A review of recent statements and press releases by OSHA indicates little, if any, change in compliance activities, although OSHA still avows more cooperation with the employer to solve safety and health problems. Where cooperation and voluntary compliance is not forthcoming, OSHA will energetically pursue enforced compliance and egregious penalties.

A comprehensive and mandatory safety and health program remains a goal of the Assistant Secretary of Labor for OSHA, Congress, and organized labor. Reform of the OSH Act remains active in Congress.

There will be continued efforts to improve targeting of unsafe workplaces, no matter their size. You can rely on continued mandatory compliance. There will be no acceptable excuse for noncompliance. But you have a buffer zone—you can also rely on the durability, flexibility, and stability of the SH&E management system that is introduced here. Review again the corporate SH&E values suggested in Chapter 3 and reaffirm your allegiance to them. They will endure and support you through whatever challenges present themselves in the future.

We say, "This is your future. You can control it and shape it just as you have in the past." You may not be able to change the conditions and circumstances of the economy, the social environment, and other external factors, but you can decide how you are going to evaluate these changes to minimize or eliminate hazards to workers. In our worldwide environment, which has changed radically since September 11, 2001, it is also evident that there are basic and fundamental survival reasons to maintain strong internal control of any process, system, or program that can contribute to or cause a negative net worth, whether monetarily, physically, or spiritually.

Be proactive—envision methods and means to optimize the changes—create your own future! Start this new SH&E management system based on values and principles. Shake everybody up, explore how you can use Internet chat rooms, the World Wide Web, or any other technological or informational advance, to enhance the safety and health of your employees. Use these innovative systems of communication, of producing goods and services, and of behavior and motivation to your advantage.

Maybe you can't easily afford the resources that might be required to accomplish the goals and objectives of the VPP criteria. Look at the new information technology and consider how it can be used to help you meet the criteria goals and objectives. For example, an enormous amount of training is required by OSHA standards *and* by the VPP criteria. This can be very expensive, unless you find ways to use worldwide and world-class information systems to provide the training for you. Start today!

In his essay, "Unimagined Futures,"[2] John Handy discusses how the organization of tomorrow will be structured, or rather, unstructured. He mentions a library in Dubrovnik, Croatia, built to tiny proportions to replace the one destroyed during the civil war a few years ago. Although the library is tiny, it is linked to *many libraries in the world* by computer. Imagine that! Vast amounts of knowledge such as this are, and will be, available to you to get your required training done.

Today, interactive computer-based training (CBT) is being administered to any number of employees. Sitting at computers, they are studying and responding on any number of topics, such as lockout/tagout, confined space entry, cumulative trauma disorder, ladder safety, housekeeping, office safety, and supervisors' safety training. They don't even have to be together in the same room. You do not have to hire an instructor, have a large classroom, buy instruction materials, or gather all of your people from wherever they are into one classroom for however long it takes—plus pay their salary, travel, lodging, and other expenses. They can be trained right in their companies' offices.

What we are saying is *use* the new stuff that is emerging in the economy, in the workplace, and in developing industries. This is another opportunity to let go of the past and look to the future for opportunities to accelerate achievement of your safety and health goals and objectives, as well as the company's goals and objectives. Turn toward the future and make it work for you. If you foresee adverse effects on your business, prepare a contingency plan to soften the impact as much as possible.

In the area of communication, which is essential to the success of VPP implementation, you will find many new information and electronic resources that can speed your messages, open channels of interactive exchange between you and your employees and among employees, and facilitate problem resolution. You do not have to call a meeting to discuss a bump or a crisis. You can communicate it instantly to everyone involved over your company network. Ask for comments, suggestions, and recommendations from everyone, and have everyone send their responses to everyone else. Summarize them and offer solutions. Get more comments. Offer one or two possible decisive options. Get comments again. Then make the decision and advise everyone what it is. Issue your policy statement on the network. All of this will take only days, perhaps even only hours, not weeks or months.

The direct communication line into your office that is one of the VPP criteria can be electronic, not telephonic. When leaders, managers, and staff open their e-mail in the morning, they will find employees' reports, complaints, suggestions, and general informative comments. They can forward these to the offices that can resolve them and request that the action office respond to them and the employee with the resolution simultaneously, within hours, not days.

The required work practice procedures and manuals to document compliance with standards, rules, and criteria can be placed on the company network for any employee to access, read, and print when needed. Maintain control of revisions and additions in the office of the safety staff or person responsible for these documents. You will avoid a tremendous amount of intensive labor and reams of paper to copy and distribute them.

There are software companies in the business of programming the best work practices and OSHA standards into computerized procedures that can be customized to your business in much less time, effort, and cost than if the employees prepare them from scratch. The customizing can be done by the employees who are most involved in doing the work covered by the procedures, as a group or a committee. There will still be unique and tailor-made work practices that can only be written by the employees doing the work. These customized compliance procedures serve as a facilitating device to achieve minimum compliance as soon as possible. Another option is to involve the employees in the development from the start. People are more apt to follow what they have developed out of their own experience than to use someone else's language and style.

Epilogue

New, improved instrumentation is being developed and is already in use daily to perform the monitoring and sampling of atmospheric and exposure conditions in your work area. Monitoring companies will come into your plant or business area, conduct the sampling, and provide a report of the results and any problems that may be present. They will also help you fix the problems if you wish.

Technology can be especially efficient in measuring the effectiveness of your SH&E performance. For example, OSHA's Program Evaluation Profile provides a matrix format that can be computerized. The result will be a database accessible by a desktop computer that will record assessments and evaluations. When the annual self-evaluation is conducted, the data is entered into the computer program. The output is a comprehensive evaluation of all aspects of the SH&E program. With this tool, you can involve all of the employees in evaluating management performance during the year and discover a measured progress toward quality and excellence.

There are limitless avenues of opportunity opening every day. Explore them. The corporate leaders and visionaries are obligated to lead the employees into the safest and highest quality of life that the future may offer.

Prepared for the Future

This book presents a set of criteria by which company leaders and managers can manage their SH&E programs. Management and leadership philosophy and techniques are applied here to the management of a model SH&E process that happens to be the OSHA Voluntary Protection Programs—already approved by jurisdictional authorities and proven by major industries in the United States. This book is based on values, because the VPP criteria is based on management and leadership principles, values, and beliefs for all workers, whatever their responsibilities, authority, and accountabilities.

We can't discuss VPP criteria without also discussing corporate culture. This book talks about how the corporate leaders, whatever their level of work, should manage the SH&E management system with as much dedication as they do the financial, personnel, and production systems, to realize the most profitable return. Principles, beliefs, and attitudes must be fundamental and genuine to attain success. Principles and values last forever.

"Principles and values last forever" is the most solid foundation on which to build corporate culture. Whatever future changes, whether they be new skills, new equipment, or new hazards that are introduced into the work environment, you will find that you are prepared for them. The SH&E management processes and systems that are now in place provide a ready means and method to derive proactive and proper preventions, controls, and protections for employees and the workplace. They are based on unchanging, natural principles—mutual trust and respect, continuous improvement, constant self-assessment and evaluation, a learning organizational culture, verifiable measurements, and open, genuine, thoughtful two-way communication with and among the employees.

We feel certain that, if you follow the recommendations and apply the techniques and methods discussed in this book, you will be prepared for "whatever whitewater comes" in the future, because you are and will remain on firm and sure footing to steer your course.

If you have ever attended one of Stephen Covey's seminars, you have heard him ask the audience if they will raise their hands and point north. In a large meeting room with no windows and no outside point of reference, it is easy to lose your sense of direction. People usually point in all directions. This brings a laugh. Then Covey takes a little compass and puts it on the projector—and on the screen you see that the needle points to magnetic north. It is our intention and hope that this book points you to the true north.

Safety, health, and the environment are about life and death—not about procedures, not about profits, not about competition—life and death. Profits and competitive advantage are only two of the many rewards for paying close, conscientious, and active attention to the control of the hazards in your company's environment for the protection of your employees.

Stay the course on true north. Principles and values last forever.

References

1. Stephen R. Covey, *Principle-Centered Leadership*. New York: Summit Books, 1991: 319–20.
2. John Handy, "Unimagined Futures," in *The Organization of the Future*. San Francisco: Josey-Bass Inc., 1997: 377–83.

Epilogue

Appendices

Appendix A
Models of Safety and Health Excellence Act of 2001

107th CONGRESS
1st Session
H. R. 2235

To authorize the Secretary of Labor to establish voluntary protection programs.

IN THE HOUSE OF REPRESENTATIVES
June 19, 2001

Mr. PETRI (for himself, Mr. ANDREWS, Mr. ISAKSON, Ms. WOOLSEY, Mr. PAUL, Mr. LAHOOD, and Mr. HUTCHINSON) introduced the following bill; which was referred to the Committee on Education and Workforce.

A BILL

To authorize the Secretary of Labor to establish voluntary protection programs.

Be it enacted by the Senate and House of Representatives of the United States of America in Congress assembled,

SEC. 1. SHORT TITLE.

This Act may be cited as the `Models of Safety and Health Excellence Act of 2001'.

SEC. 2. FINDINGS AND PURPOSES.

(a) FINDINGS- The Congress finds the following:

(1) Since 1982, the Occupational Safety and Health Administration has conducted voluntary protection programs designed to recognize excellence in occupational safety and health.

(2) Such programs have fostered partnerships between employers, employees, and the Occupational Safety and Health Administration to improve workplace safety and health through the implementation of effective safety and health programs.

(3) Employers participating in such programs provide their employees with a level of protection that substantially exceeds the level of protection provided by compliance with the requirements of the Occupational Safety and Health Act of 1970 (29 U.S.C. 651 et seq.).

(4) As a result of these efforts, employers participating in such programs have experienced injury and illness rates that are on average less than half of their respective industry averages, sparing thousands of America's working families needless workplace tragedies.

(e) PURPOSES—The purposes of this Act are the following:

(1) To recognize the exemplary leadership of voluntary protection programs participants in improving occupational safety and health at workplaces;

(2) To encourage other employers to adopt such approaches to protect their workers; and

(3) To codify such programs to ensure that the Occupational Safety and Health Administration continues to develop them in the future.

SEC. 3. VOLUNTARY PROTECTION PROGRAMS.

(a) IN GENERAL—The Secretary of Labor or the Secretary's authorized representative shall establish and carry out voluntary protection programs (hereinafter in this section referred to as `programs') to promote and recognize the achievement of worksites that demonstrate excellence in workplace health and safety. The Secretary may choose, in limited situations, to alter the application requirements in order to expand the scope of worksites participating in the programs to include nonstandard worksites such as short-term construction sites and mobile worksites. The Secretary shall encourage the participants in the programs to share occupational safety and health expertise with other employers. The Secretary shall also encourage the participation of small business (as that term is defined by the Administrator of the Small Business Administration) in the programs by implementing outreach and assistance initiatives in cooperation with program participants and shall develop program requirements that address the needs of small businesses. The Secretary may provide for the development of equivalent programs in State plan States.

(b) PROGRAM REQUIREMENTS—A program shall include the following:

(1) APPLICATION—Applications for participation in the programs shall be submitted by the worksite's management, shall reflect the support of a substantial number of site employees, and, where appli-

cable, shall have the express written support of the collective bargaining representative of such employees. Employers who volunteer under the programs shall be required to submit an application to the Secretary of Labor demonstrating that the worksite, with respect to which the application is made, meets such requirements as the Secretary may require for participation in the program. Such requirements shall include demonstrations of exemplary comprehensive programs to assure—

  (A) upper management leadership and active and meaningful employee involvement;

  (B) systematic assessment of hazards;

  (C) comprehensive hazard prevention, mitigation, and control programs;

  (D) employee safety and health training; and

  (E) safety and health program evaluation.

(6) ONSITE EVALUATIONS—There shall be onsite evaluations of each permanent worksite by representatives of the Secretary and others from the private and public sector as determined by the Secretary.

(7) INFORMATION—Employers who are approved by the Secretary for participation in a program shall assure the Secretary that such information as is necessary to evaluate the employer's application and continued participation in the program is made available to the Secretary.

(8) REEVALUATIONS—Periodic reevaluations by the Secretary shall be required for continued participation in a program.

(i) PROGRAM ADMINISTRATION—

(1) EXEMPTIONS—A worksite which has been selected to participate in a program shall, while participating in the program, be exempt from inspections or investigations under the Occupational Safety and Health Act of 1970 (29 U.S.C. 651 et seq.), except that the exemption shall not apply to inspections or investigations arising from employee complaints, fatalities, catastrophes, or significant toxic releases.

(2) PROGRAM ACCEPTANCE AND CONTINUED PARTICIPATION—Decisions regarding acceptance into the program and continued participation in the program will be based on the

Appendix A

applicant's superior safety and health performance, as determined by the Secretary or the Secretary's authorized representatives.

(3) PROGRAM PARTICIPATION- Decisions regarding participation in a program are in the sole discretion of the Secretary or an authorized representative of the Secretary.

END

Appendix B
Sources of Help

Advisors, Coaches, and Mentors

Members of the Voluntary Protection Programs Participants' Association (VPPPA) (see listing below) are encouraged to mentor, coach, and advise any employer who is interested in improving his or her safety and health management system.

In the OSHA Directive TED 8.4, Appendix B, OSHA defines the "Mentoring Program" as follows:

III. Mentoring

- A. Purpose: The goal of the mentoring program is to use proven skills found among employees at VPP sites to help protect workers at other sites through implementation or improvement of the safety and health management systems. The program is voluntary and coordinated from the National Office of the VPPPA in Falls Church, Virginia.

- B. Background: The Mentoring Program was established in Spring 1994 by the VPPPA in consultation with OSHA. The program formalizes and expands the mentoring and assistance that VPP participants were providing to facilities interested in pursuing application to OSHA's VPP. The scope of the program has since been expanded to assist applicants to the DOE's VPP and any site that wants to improve its safety and health management system. Mentors provide guidance in developing the VPP application and share techniques to improve a facility's safety and health management system.

American Society for Quality

600 North Plankinton Avenue
PO Box 3005
Milwaukee, WI 53201-3005

Telephone: 800 248 1946
Fax: 414 272 1734
Web: www.asq.org

The American Society for Quality (ASQ) is a society of individual and organizational members dedicated to the ongoing development, advancement, and promotion of quality concepts, principles, and techniques. ASQ administers the Malcolm Baldrige National Quality Award Program under contract to NIST.

American Society for Quality Control

611 East Wisconsin Avenue
PO Box 3066
Milwaukee, WI 53201-3066

Telephone: 800 248 1946
Fax: 414 272 1734
Web: www.asq.org

The American Society for Quality Control (ASQC) is the US member of the American National Standards Institute (ANSI) responsible for quality management and related standards. As ANSI's representative, the ASQC administers the quality management and related standards issued by the International Organization for Standardization (ISO), of which ANSI is the US member. The mission of the ASQC is to "facilitate continuous improvement and increase customer satisfaction by identifying, communicating, and promoting the use of quality principles, concepts, and technologies; and thereby be recognized throughout the world as the leading authority on, and champion for, quality."

The National Institute of Standards and Technology (NIST)

United States Department of Commerce
Technology Administration
National Institute of Standards and Technology
National Quality Program
Route 270 and Quince Orchard Road
Administration Building, Room A635
Gaithersburg, MD 20899-0001

Telephone: 301 975 2036
Fax: 301 948 3716
Web: www.nist.gov

The National Institute of Standards and Technology (NIST) is a nonregulatory federal agency within the Commerce Department's Technology Administration. NIST's primary mission is to promote economic growth by working with industry to develop and apply technology, measurements, and standards. Call NIST for:

- Information about the Criteria for Performance Excellence
- Individual copies of the Criteria (no cost)
- Application forms and instructions (no cost)
- Examiner applications (no cost)

Occupational Safety and Health Administration (OSHA)

(National OSHA office for the VPP)
Office of Cooperative Programs
OSHA
200 Constitution Ave NW
Washington, DC 20210
Web: www.osha.com

In the ten regions of OSHA, there are VPP managers or coordinators whose assigned duties are to assist employers who already have been accepted into the programs or who are interested in improving their safety and health programs. The OSHA regional office is listed in your phone book or the local OSHA office can provide the number.

Some local offices are now conducting VPP site evaluations, as well as compliance visits. Contact the OSHA office in your area to get more information.

There is a world of information regarding standards, compliance, OSHA activities, and VPP on the OSHA website.

Voluntary Protection Programs Participants' Association (VPPPA)

Voluntary Protection Programs Participants' Association
7600 E. Leesburg Pike, Suite 440
Falls Church, VA. 22043-2004
Paul M. Villane, CSP, OHST
Executive Director

Telephone: 703 761 1146
Fax: 703 761 1148
Web: www.vpppa.org

The VPPPA, a national nonprofit organization, was formed in 1985 and is known by key players in workplace safety, health, and environmental excellence as a "one-stop resource." The VPPPA represents the 1,000-plus worksites that are either participating in, or have applied to participate in, OSHA's or the Department of Energy's VPP, as well as the government agencies that administer these cooperative programs. The VPPPA can offer mentoring services that match interested facilities with mentors who share safety and health information and provide assistance.

The VPPPA offers courses and workshops throughout the US tailored expressly for those involved or who want to be involved in cooperative programs such as the VPP. The courses and workshops are geared to sites that want to take a proactive approach to safety and health and are implementing a safety and health program.

Held in a different US city each year, the VPPPA national conference brings together more than 2,000 governmental officials, safety, health, and environmental hourly workers and managers from all over the country. More than 100 workshops, exhibits, and special sessions are featured. The attendees gain valuable knowledge in site reviews, ergonomics, environmental safety, management commitment, and the most current OSHA safety and health requirements.

The VPPPA has several tools available to inform members and subscribers of the latest information on cooperative programs and best practices in safety, health, and the environment. *The Leader* provides the latest in safety, health, and environmental issues, breaking news, and achievements of VPPPA member companies and their employees. Interested readers can contact the addresses above to subscribe to the publication or obtain other information.

Other Websites

The following websites offer relevant information on safety, health, and environmental issues.

American Industrial Hygiene Association	www.aiha.org
American Society of Safety Engineers	www.asse.org
Association of Reciprocal Safety Councils	www.arsc.com
National Safety Council	www.nsc.org
US Department of Energy	www.energy.gov
US Environmental Protection Agency	www.epa.gov

Appendix C

Requirements for Star, Merit, Resident Contractor, Construction Industry, and Federal Agency Worksites

Verbatim extract from OSHA Instruction Directive Number TED 8.4, *Voluntary Protection Programs (VPP): Policies and Procedures Manual,* effective date: March 25, 2003

Chapter III
Requirements for Star, Merit, Resident Contractor, Construction Industry, and Federal Agency Worksites

I. **Introduction.** This chapter delineates requirements for the Star and Merit programs as well as the unique requirements for the construction industry, for Federal agencies, and for resident contractors at VPP sites.

II. **The Star Program.** The Star Program recognizes the very best workplaces that are in compliance with OSHA regulations and that operate outstanding safety and health management systems for worker protection. All of the VPP requirements, published in *Federal Register* Notice 65 FR 45650-45663 and detailed below, must be in place and working effectively for at least 1 year prior to Star approval.

   A. **Term of Participation.** There is no limit to the term of participation in Star, as long as a site continues to meet all Star requirements and to maintain Star quality.

   B. **Injury and Illness History Requirements.** Injury and illness history at the site is evaluated using a 3-year total case incident rate (TCIR) and a 3-year day away, restricted, and/or transfer case incident rate (DART rate). (See Appendix A.) The 3-year TCIR and DART rates must be compared to the most recently published Bureau of Labor Statistics (BLS) national average for the three or four-digit (if available) Standard Industrial Classification code (SIC). The TCIR and DART rates must be compared to the five- or six-digit North American Industrial Classification System (NAICS) code for the industry in which the applicant is classified when the NAICS system is adopted. The BLS publishes SIC and NAICS industry averages 2 years after data is collected. (For example, in calendar year 2003, calendar year 2001 national averages will be available and used for comparison).

1. Both the 3-year TCIR and the 3-year DART rate must be below the most recently published BLS national average for the specific SIC or NAICS code.

2. Some smaller worksites may be eligible to use the best 3 out of the most recent 4 years for calculating incidence rates. (See Appendix A, TED 8.4.)

3. Transition to OSHA 300 logs:

   VPP will institute the following policy until such time as three consecutive years of recording data that utilizes the new record-keeping standard has been published. Until that time, and to address the variability that is expected in the industrial accident rate averages, the following policy will be in effect:

   The requirement for a Star site to have a 3-year injury and illness rate that is below their respective industry is to be modified such that an applicant/participant's rate must be below their respective industries injury and illness rate plus 10%.

   As a result a Star worksite may have an injury and illness rate above the national average if it is within 10% of that BLS average.

C. **Comprehensive Safety and Health Management System Requirements.** The following safety and health management system elements and sub-elements must be implemented. For small sites, at the discretion of the onsite team, some of the requirements may be implemented and documented less formally.

   1. Management Leadership and Employee Involvement

      a. **Management Commitment.** Management demonstrates its commitment by:

         - Establishing, documenting, and communicating to employees and contractors clear goals that are attainable and measurable, objectives that are relevant to workplace hazards and trends of injury and illness, and policies and procedures that indicate how to accomplish the objectives and meet the goals.
         - Signing a statement of commitment to safety and health.
         - Meeting and maintaining VPP requirements.

- Maintaining a written safety and health management system that documents the elements and sub-elements, procedures for implementing the elements, and other safety and health programs including those required by OSHA standards.
- Identifying persons whose responsibilities for safety and health includes carrying out safety and health goals and objectives, and clearly defining and communicating their responsibilities in their written job descriptions.
- Assigning adequate authority to those persons who are responsible for safety and health, so they are able to carry out their responsibilities.
- Providing and directing adequate resources (including time, funding, training, personnel, etc.) to those responsible for safety and health, so they are able to carry out their responsibilities.
- Holding those assigned responsibility for safety and health accountable for meeting their responsibilities through a documented performance standards and appraisal system.
- Planning for typical as well as unusual/emergency safety and health expenditures in the budget, including funding for prompt correction of uncontrolled hazards.
- Integrating safety and health into other aspects of planning, such as planning for new equipment, processes, buildings, etc.
- Establishing lines of communication with employees and allowing for reasonable employee access to top management at the site.
- Setting an example by following the rules, wearing any required personal protective equipment, reporting hazards, reporting injuries and illnesses, and basically doing anything that they expect employees to do.
- Ensuring that all workers (including contract workers) are provided equal, high-quality safety and health protection.

- Conducting an annual evaluation of the safety and health management system in order to:
  - Maintain knowledge of the hazards of the site.
  - Maintain knowledge of the effectiveness of system elements.
  - Ensure completion of the previous years' recommendations.
  - Modify goals, policies, and procedures.

b. **Employee Involvement.** Employees must be involved in the safety and health management system in at least three meaningful, constructive ways in addition to their right to report a hazard. Avenues for employees to have input into safety and health decisions include participation in audits, accident/incident investigations, self-inspections, suggestion programs, planning, training, job hazard analyses, and appropriate safety and health committees and teams. Employees do not meet this requirement by participating in incentive programs or simply working in a safe manner.

- Employees must be trained for the task(s) they will perform. For example, they must be trained in hazard recognition to participate in self-inspections.
- Employees must receive feedback on any suggestions, ideas, reports of hazards, etc. that they bring to management's attention. A site must provide documented evidence that employees' suggestions were followed up and implemented when appropriate and feasible.
- All employees, including new hires, must be notified about the site's participation in VPP and employees' rights (such as the right to file a complaint) under the OSH Act. Orientation training curriculum must include this information.
- Employees and contractors must demonstrate an understanding of and be able to describe the fundamental principles of VPP.

c. **Contract Worker Coverage.** Contract workers must be provided with safety and health protection equal in quality to that provided to employees.

- All contractors, whether regularly involved in routine site operations or engaged in temporary projects such as construction or repair, must follow the safety and health rules of the host site.

- VPP participants must have in place a documented oversight and management system covering applicable contractors. Such a system must:
  - Ensure that safety and health considerations are addressed during the process of selecting contractors and when contractors are onsite.
  - Encourage contractors to develop and operate effective safety and health management systems.
  - Include provisions for timely identification, correction, and tracking of uncontrolled hazards in contractor work areas.
  - Include a provision for removing a contractor or contractor's employees from the site for safety or health violations. Note: A site may have been operating effectively for 1 year without actually invoking this provision if just cause to remove a contractor or contractor's employee did not occur.
- Injury and Illness Data Requirements
  - Nested contractors (such as contracted maintenance workers) and temporary employees who are supervised by host site management are governed by the site's safety and health management system and are therefore included in the host site's rates.
  - Site management must maintain copies of the TCIR and DART rate data for all applicable contractors based on hours worked at the site. (See Appendix A, TED 8.4.)
  - Sites must report all applicable contractors' TCIR and DART rate data to OSHA annually.
- **Training.** Managers, supervisors, and non-supervisory employees of contract employers must be made aware of:
  - The hazards they may encounter while on the

Appendix C

site.

- How to recognize hazardous conditions and the signs and symptoms of workplace-related illnesses and injuries.

- The implemented hazard controls, including safe work procedures.

- Emergency procedures.

d. **Safety and Health Management System Annual Evaluation.** There must be a system and written procedures in place to annually evaluate the safety and health management system. The annual evaluation must be a critical review and assessment of the effectiveness of all elements and sub-elements of a comprehensive safety and health management system. An annual evaluation that is merely a workplace inspection with a brief report pointing out hazards or a general statement of the sufficiency of the system is inadequate for purposes of VPP qualification.

- The written annual evaluation must identify the strengths and weaknesses of the safety and health management system and must contain specific recommendations, time lines, and assignment of responsibility for making improvements. It must also document actions taken to satisfy the recommendations.

- The annual evaluation may be conducted by site employees with managers, qualified corporate staff, or outside sources who are trained in conducting such evaluations.

- At least one annual evaluation and demonstrated corrective action must be completed before VPP approval.

- The annual evaluation must be included with the participant's annual submission to OSHA. Appendix D provides a suggested format.

2. **Worksite Analysis.** A hazard identification and analysis system must be implemented to systematically identify basic and unforeseen safety and health hazards, evaluate their risks, and prioritize and recommend methods to eliminate or control hazards to an acceptable level of risk. Through this system, management must gain a thorough knowledge of the safety and health hazards and employee risks. The required methods of hazard identification and analysis are described below.

a. **Baseline Safety and Industrial Hygiene Hazard Analysis.** A baseline survey and analysis is a first attempt at understanding the hazards at a worksite. It establishes initial levels of exposure (baselines) for comparison to future levels, so that changes can be recognized. Systems for identifying safety and industrial hygiene hazards, while often integrated, may be evaluated separately. Baseline surveys must:

- Identify and document common safety hazards associated with the site (such as those found in OSHA regulations or building standards, for which existing controls are well known), and how they are controlled.

- Identify and document common health hazards (usually by initial screening using direct-reading instruments) and determine if further sampling (such as full-shift dosimetry) is needed.

- Identify and document safety and health hazards that need further study.

- Cover the entire work site, indicate who conducted the survey, and when it was completed.

The original baseline hazard analysis need not be repeated subsequently unless warranted by changes in processes, equipment, hazard controls, etc.

b. **Hazard Analysis of Routine Jobs, Tasks, and Processes.** Task-based or system/process hazard analyses must be performed to identify hazards of routine jobs, tasks, and processes in order to recommend adequate hazard controls. Acceptable techniques include, but are not limited to: Job Hazard Analysis (JHA), and Process Hazard Analysis (PHA).

Hazard analyses should be conducted on routine jobs, tasks and processes that:

- Have written procedures.

- Have had injuries/illnesses associated with them or have experienced significant incidents or near-misses.

- Are perceived as high-hazard tasks, i.e., they could result in a catastrophic explosion, electrocution, or chemical over-exposure.

- Have been recommended by other studies and analyses for more in-depth analysis.
- Are required by a regulation or standard.
- Any other instance when the VPP applicant or participant determines that hazard analysis is warranted.

c. **Hazard Analysis of Significant Changes.** Hazard analysis of significant changes, including but not limited to, non-routine tasks (such as those performed less than once a year), new processes, materials, equipment and facilities must be conducted to identify uncontrolled hazards prior to the activity or use, and must lead to hazard elimination or control.

If a non-routine or new task is eventually to be done on a routine basis, then a hazard analysis of this routine task should subsequently be developed.

d. **Pre-use analysis.** When a site is considering new equipment, chemicals, facilities, or significantly different operations or procedures, the safety and health impact to the employees must be reviewed. The level of detail of the analysis should be commensurate with the perceived risk and number of employees affected. This practice should be integrated in the procurement/design phase to maximize the opportunity for proactive hazard controls.

e. **Documentation and Use of Hazard Analyses.** Hazard analyses performed to meet the requirements of c. or d. above must be documented and must:

- Consider both health and safety hazards.
- Identify the steps of the task or procedure being analyzed, hazard controls currently in place, recommendations for needed additional or more effective hazard controls, dates conducted, and responsible parties.
- Be used in training in safe job procedures, in modifying workstations, equipment or materials, and in future planning efforts.
- Be easily understood.
- Be updated as the environment, procedures, or equipment change, or errors are found that invalidate the most recent hazard analyses.

f. **Routine Self-Inspections.** A system is required to ensure routinely scheduled self-inspections of the workplace. It must include written procedures that determine the frequency of inspection and areas covered, those responsible for conducting the inspections, recording of findings, responsibility for abatement, and tracking of identified hazards for timely correction. Findings and corrections must be documented.

- Inspections must be made at least monthly, with the actual inspection schedule being determined by the types and severity of hazards.
- The entire worksite must be covered at least once each quarter.
- Top management and others, including employees who have knowledge of the written procedures and hazard recognition, may participate in the inspection process.
- Personnel qualified to recognize workplace hazards, particularly hazards peculiar to their industry, must conduct inspections.
- Documentation of inspections must evidence thoroughness beyond the perfunctory use of checklists.

g. **Hazard Reporting System for Employees.** The site must operate a reliable system that enables employees to notify appropriate management personnel in writing—without fear of reprisal—about conditions that appear hazardous and to receive timely and appropriate responses. The system can be anonymous and must include timely responses to employees and tracking of hazard elimination or control to completion.

h. **Industrial Hygiene (IH) Program.** A written IH program is required. The program must establish procedures and methods for identification, analysis, and control of health hazards for prevention of occupational disease.

- **IH Surveys.** Additional expertise, time, technical equipment, and analysis beyond the baseline survey may be required to determine which environmental contaminants (whether physical, biological, or chemical) are present in the workplace and to quantify exposure so that proper controls can be implemented.

Appendix C

- **Sampling Strategy.** The written program must address sampling protocols and methods implemented to accurately assess employees' exposure to health hazards. Sampling should be conducted when:
  - Performing baseline hazard analysis, such as initial screening and grab sampling.
  - Baseline hazard analysis suggests that more in-depth exposure analysis, such as full-shift sampling, is needed.
  - Particularly hazardous substances (as indicated by an OSHA standard, chemical inventory, material safety data sheet, etc.) are being used or could be generated by the work process.
  - Employees have complained of signs of illness.
  - Exposure incidents or near-misses have occurred.
  - It is required by a standard or other legal requirement.
  - Changes have occurred in such things as the processes, equipment, or chemicals used.
  - Controls have been implemented and their effectiveness needs to be determined.
  - Any other instance when the VPP applicant or participant determines that sampling warranted.
- **Sampling Results.** Sampling results must be analyzed and compared to at least OSHA permissible exposure limits (PELs) to determine employees' exposure and possible overexposure. Comparison to more restrictive levels, such as action levels, threshold limit values (TLVs), or self-imposed standards is encouraged to reduce exposures to the lowest feasible level.
- **Documentation.** The results of sampling must be documented and must include a description of the work process, controls in place, sampling time, exposure calculations, duration, route, and frequency of exposure, and number of exposed employees.
- **Communication.** Sampling results must be communicated to employees and management.
- **Use of Results.** Sampling results must be used to identify

areas for additional, more in-depth study, to select hazard controls, and to determine if existing controls are adequate.

- **IH Expertise**. IH sampling should be performed by an industrial hygienist, but initial sampling, full-shift sampling, or both may be performed by safety staff members with special training in the specific procedures for the suspected or identified health hazards in the workplace.

- **Procedures**. Standard, nationally recognized procedures must be used for surveying and sampling as well as for testing and analysis.

- **Use of Contractors**. If an outside contractor conducts industrial hygiene surveys, the contractor's report must include all sampling information listed above and must be effectively communicated to site management. Any recommendations contained in the report should be considered and implemented where appropriate. Use of contractors does not remove responsibility for the IH program, including identification and control of health hazards, from the VPP applicant or participant.

i. **Investigation of Accidents and Near-Misses**. The site must investigate all accidents and near-misses and must maintain written reports of the investigations. Accident and near-miss investigations must:

- Be conducted by personnel trained in accident investigation techniques. Personnel who were not involved in the accident or who do not supervise the injured employee(s) should conduct the investigation to minimize potential conflicts of interest.

- Document the entire sequence of relevant events.

- Identify all contributing factors, emphasizing failure or lack of hazard controls.

- Determine whether the safety and health management system was effective, and where it was not provide recommendations to prevent recurrence.

- Not place undue blame or reprisal on employees, although human error can be a contributing factor.

- Assign priority, time frames, and responsibility for implementing recommended controls.
- The results of investigations (to include, at a minimum, a description of the incident and the corrections made to avoid recurrence) must be made available to employees on request, although the actual investigation records need not be provided.

j. **Trend Analysis**. The process must include analysis of information such as injury/illness history, hazards identified during inspections, employee reports of hazards, and accident and near-miss investigations for the purpose of detecting trends. The results of trend analysis must be shared with employees and management and utilized to direct resources; prioritize hazard controls; and determine or modify goals, objectives, and training to address the trends.

3. **Hazard Prevention and Control**. Management must ensure the effective implementation of systems for hazard prevention and control and ensure that necessary resources are available, including the following:

   a. **Certified Professional Resources**. Access to certified safety and health professionals and other licensed health care professionals is required. They may be provided by offsite sources such as corporate headquarters, insurance companies, or private contractors. OSHA will accept certification from any recognized accrediting organization.

   b. **Hazard Elimination and Control Methods**. The types of hazards employees are exposed to, the severity of the hazards, and the risk the hazards pose to employees should all be considered in determining methods of hazard prevention, elimination, and control. In general, the following hierarchy should be followed in determining hazard elimination and control methods. When engineering controls have been studied, investigated, and implemented, yet still do not bring employees' exposure levels to below OSHA permissible exposure limits; or when engineering controls are determined to be infeasible, then a combination of controls may be used. Whichever controls a site chooses to employ, the controls must be understood and followed by all affected parties; appropriate to the site's hazards; equitably enforced through the disciplinary system; written, implemented, and updated by management as needed; used by employees; and incorporated in training, positive reinforcement, and correction programs.

- Engineering. Engineering controls directly eliminate a hazard by such means as substituting a less hazardous substance, by isolating the hazard, or by ventilating the workspace. These are the most reliable and effective controls.

- Protective Safety Devices. Although not as reliable as true engineering controls, such methods include interlocks, redundancy, fail-safe design, system protection, fire suppression, and warning and caution notes.

- Administrative. Administrative controls significantly limit daily exposure to hazards by control or manipulation of the work schedule or work habits. Job rotation is a type of administrative control.

- Work Practices. These controls include workplace rules, safe and healthful work practices, personal hygiene, housekeeping and maintenance, and procedures for specific operations.

- Personal Protective Equipment (PPE). PPE to be used are determined by hazards identified in hazard analysis. PPE should only be used when all other hazard controls have been exhausted or more significant hazard controls are not feasible.

c. **Hazard Control Programs**. Applicants and participants must be in compliance with any hazard control program required by an OSHA standard, such as PPE, Respiratory Protection, Lockout/Tagout, Confined Space Entry, Process Safety Management, or Bloodborne Pathogens. VPP applicants and participants must periodically review these programs (most OSHA standards require an annual review) to ensure they are up to date.

d. **Occupational Health Care Program**

- Licensed health care professionals must be available to assess employee health status for prevention, early recognition, and treatment of illness and injury.

- Arrangements for needed health services such as pre-placement physicals, audiograms, and lung function

Appendix C

tests must be included.

- Employees trained in first aid, CPR providers, physician care, and emergency medical care must be available for all shifts within a reasonable time and distance. The applicant or participant may consider, based on site conditions, providing Automated External Defibrillators (AEDs) and training in their use.

- Emergency procedures and services including provisions for ambulances, emergency medical technicians, emergency clinics or hospital emergency rooms should be available and explained to employees on all shifts. Also see paragraph h below.

e. **Preventive Maintenance of Equipment.** A written preventive and predictive maintenance system must be in place for monitoring and maintaining workplace equipment. Equipment must be replaced or repaired on a schedule, following manufacturers' recommendations, to prevent it from failing and creating a hazard. Documented records of maintenance and repairs must be kept. The system must include maintenance of hazard controls such as machine guards, exhaust ventilation, mufflers, etc.

f. **Tracking of Hazard Correction.** A documented system must be in place to ensure that hazards identified by any means (self-inspections, accident investigations, employee hazard reports, preventive maintenance, injury/illness trends, etc.) are assigned to a responsible party and corrected in a timely fashion. This system must include methods for:

- Recording and prioritizing hazards, and

- Assigning responsibility, time-frames for correction, interim protection, and follow-up to ensure abatement.

g. **Disciplinary System.** A documented disciplinary system must be in place. The system must include enforcement of appropriate action for violations of the safety and health policies, procedures, and rules. The disciplinary policy must be clearly communicated and equitably enforced to employees and management. The disciplinary system for safety and health can be a sub-part of an all-encompassing disciplinary system.

h. **Emergency Preparedness and Response.** Written procedures for response to all types of emergencies (fire, chemical spill, accident, terrorist threat, natural disaster, etc.) on all shifts must be established, must follow OSHA standards, must be communicated to all employees, and must be practiced at least annually. These procedures must list requirements or provisions for:

- Assessment of the emergency.
- Assignment of responsibilities (such as incident commander).
- First aid.
- Medical care.
- Routine and emergency exits.
- Emergency telephone numbers.
- Emergency meeting places.
- Training drills, minimally including annual evacuation drills. Drills must be conducted at times appropriate to the performance of work so as not to create additional hazards. Coverage of critical operations must be provided so that all employees have an opportunity to participate in evacuation drills.
- Documentation and critique of evacuation drills and recommendations for improvement.
- Personal protective equipment where needed.

4. **Safety and Health Training**

   a. Training must be provided so that managers, supervisors, non-supervisory employees, and contractors are knowledgeable of the hazards in the workplace, how to recognize hazardous conditions, signs and symptoms of workplace-related illnesses, and safe work procedures.

   b. Training required by OSHA standards must be provided in accordance with the particular standard.

   c. Managers and supervisors must understand their safety and health responsibilities and how to carry them out effectively.

   d. New employee orientation/training must include, at a minimum, discussion of hazards at the site, protective measures, emergency evacuation, employee rights under the OSH Act, and VPP.

e. Training should be provided for all employees regarding their responsibilities for each type of emergency. Managers, supervisors, and non-supervisory employees, including contractors and visitors, must understand what to do in emergency situations.

f. Persons responsible for conducting hazard analysis, including self-inspections, accident./incident investigations, job hazard analysis, etc., must receive training to carry out these responsibilities, e.g., hazard recognition training, accident investigation techniques, etc.

g. Training attendance must be documented. Training frequency must meet OSHA standards or, for non-OSHA required training, be provided at adequate intervals. Additional training must be provided when in-work processes, new equipment, new procedures, etc. occur.

h. Training curricula must be up-to-date, specific to worksite operations, and modified when needed to reflect changes and/or new workplace procedures, trends, hazards and controls identified by hazard analysis. Training curricula must be understandable for all employees.

i. Persons who have specific knowledge or expertise in the subject area must conduct training.

j. Where personal protective equipment (PPE) is required, employees must understand that it is required, why it is required, its limitations, how to use it, and maintenance.

III. **Merit Program.** The Merit program recognizes worksites that have good safety and health management systems but must take additional steps to reach Star quality. If OSHA determines that an employer has demonstrated the commitment and possesses the resources to meet Star requirements within 3 years, the employer may enter the Merit program with set goals for reaching Star.

   A. **Injury and Illness History Requirements.** The TCIR and DART rate must be calculated and compared to the industry average in the same manner as for the Star Program, except that the 3-year rates do not have to be below the industry average. The following restrictions apply:

   1. If the site has either or both the TCIR and DART rate above the industry average, the site must set realistic, concrete goals for reducing both rates within 2 years and must specify the methods (approved by the VPP Manager) to be used to accomplish the

goals.

2. It must be programmatically and statistically feasible for the site to reduce its TCIR and DART rate to below the industry average within 2 years.

B. **Comprehensive Safety and Health Management System Requirements.** The basic elements and sub-elements described for Star participation (Management Leadership and Employee Involvement, Worksite Analysis, Hazard Prevention and Control, Safety and Health Training) must all be operational or, at a minimum, in place and ready for implementation by the date of approval. In addition, all minimum requirements (MRs) must be met. (See Appendix D, TED 8.4, Appendix E, in this book.)

C. **Merit Goals.** If the Onsite evaluation team recommends participation in the Merit program, the site must then complete a set of goals in order to maintain Merit status and qualify for the Star Program.

1. Merit goals must address Star requirements not presently in place or aspects of the safety and health management system that are not up to Star quality.

2. Methods for improving the safety and health management system that will address identified problem areas must be included in Merit goals.

3. Correction of a specific hazardous condition must be a 90-day item, not a Merit goal. However, when a safety and health management system deficiency underlies a specific hazardous condition, then corrections to the system must be included as Merit goals.

4. Reducing a 3-year TCIR or DART rate to below the national average is not, by itself, an appropriate Merit goal. Corrections to safety and health management system deficiencies underlying the high rate must be included in the Merit goals.

D. **Term of Participation.** The length of term is dependent on the time necessary to accomplish Merit goals; however, initial approval to Merit will be for a single term not to exceed 3 years. [See Chapter 6.VIII.C.2, TED 8.4.]

1. A site must meet Star rate requirements within the first 2 years of its Merit participation. This is to afford an additional year's experience, for a total of no more than 3 years to gain Star approval.

Appendix C

2. A Merit sites qualifies for Star when it has met its Merit goals, Star rate requirements, and when all other safety and health elements and sub-elements are operating at Star quality.

3. A Merit site may qualify for the Star Program before the end of its Merit term if the site meets all conditions in 2. above.

IV. **Resident Contractors.** Contractors working at a VPP site may apply to VPP. The requirements for the resident contractor are identical to those of VPP generally, with the following additions:

A. The host site must be an approved VPP (Star or Merit) site before the resident contractor may submit its application. In addition, the resident contractor must have a minimum of 12 months on site before submitting an application.

B. The type of work being conducted by the resident contractor must be evaluated to determine the appropriate industry classification.

1. If the resident contractor is fulfilling a function that would normally be filled by the host (such as general maintenance), then the resident contractor should be assigned the host's industry classification.

2. If the resident contractor is independent and would not normally be associated with the host site's industry or service, then the contractor's own industry classification should be assigned.

C. If the resident contractor has less than 3 years on site, apply the injury and illness history requirements for construction. [See V.A. below.]

D. The resident contractor's participation, once approved, is contingent upon the host site's continued participation in VPP.

E. A general contractor (GC) of a large construction project at an approved VPP site can submit a separate application for VPP. The requirements for construction apply.

F. **Replacement of an Approved Resident Contractor.** When a VPP resident contractor at a VPP site is replaced by a new resident contractor, whether VPP approval will transfer to the new resident contractor depends on several factors.

1. VPP status can transfer if 75% or more of the employees remain employed with the new resident contractor and if the new resident contractor:

a. Submits a new letter of management commitment,

b. Submits a new self-evaluation, and

        c. Receives a satisfactory OSHA onsite evaluation within 12 months (6 months is preferred).
   2. A new VPP application is required if fewer than 75% of the employees remain employed with the new resident contractor.
   G. **Continuing VPP status of an approved subcontractor** to the initial resident contractor depends on the status of the new resident contractor.
      1. If VPP status transfers to the new resident contractor, as in F1.above, the subcontractor maintains its VPP status.
      2. If the new resident contractor is required to submit a new VPP application, as in F2.above, the subcontractor must withdraw from VPP and then reapply after approval of the new resident contractor.

V. **VPP Requirements for the Construction Industry** A construction applicant must be the general contractor (GC), owner, or an organization that provides overall management at a site, controls site operations, and has ultimate responsibility for assuring safe and healthful working conditions at the site. The project must have been in operation for at least 12 months prior to approval. Construction applications cover individual sites only.

   A. Injury/Illness History
      1. To qualify for the Star Program, the site's TCIR and DART rate (including all subcontractor workers at the site) from site inception until time of application must be below the national average for the industry classification.
      2. If a site has rates which exceed the BLS average for it's SIC, then the general contractor may qualify for the Merit program if the company-wide 3-year TCIR and DART rates are below the national average.

      The site may use nationwide employment data, or may designate, with OSHA approval, an appropriate geographical area to determine employee coverage.

   B. **Comprehensive Safety and Health Management System Requirements.** The requirements for the Star and Merit programs are identical to those of VPP generally, with the following additions:
      1. **Safety and Health Management System Evaluation.** The

evaluation must be conducted annually and immediately prior to completion of construction. If a construction company does not provide the final evaluation, OSHA will not consider subsequent VPP applications for other sites operated by that company.

2. **Routine Self-Inspections.** These inspections must cover the entire worksite at least weekly, due to the changing nature of construction sites.

3. The applicant or participant is responsible for ensuring the correction of any identified hazards, including those created by subcontractors.

4. General Contractors must make subcontractors and their employees aware of the VPP application or participation and of their rights, roles and responsibilities. Evidence that all subcontractors at the site recognize these conditions is necessary and may include:

   a. The contractual agreement.

   b. A written statement of willingness to cooperate.

   c. Attendance at safety meetings.

   d. Orientation sessions for incoming subcontractor employees.

5. **Employee Involvement.** Employees at construction sites must be involved in safety and health at the site to the degree practical based on the time they will spend on the site. Examples of short-term involvement include attending daily toolbox talks on safety and health, and participating in daily self-inspections. The more time they spend on site, the more involvement OSHA expects. The onsite evaluation team will judge the sufficiency of employee involvement through interviews and observations.

VI. **VPP for Federal Agencies** Federal agency worksites that are covered by 29 CFR 1960 and fall under OSHA's jurisdiction are eligible to participate in the VPP.

   A. Requirements for Federal agencies are identical to those of VPP generally, with the following additions:

   1. In addition to complying with 29 CFR 1910 or 1926, Federal agency worksites must also be in full compliance with 29 CFR 1960.

   2. Federal worksites must notify their Designated Agency Safety and Health Official (DASHO) in writing of their intent to apply to the VPP and must submit a copy of that written notification with their application.

3. Federal worksites must make available to the OSHA onsite evaluation team their agency's current Annual Occupational Safety and Health Report to the Secretary of Labor. Any applicable elements should be noted and corrected.
4. Federal worksites must make available to the OSHA onsite evaluation team copies of their OSHA 200/300 logs and Federal Occupational Injury and Illness Log for the most recent 3 calendar years.
5. Federal worksites must not have any open Notices.
6. Federal worksites may be subject to strict contract language and regulations governing selection and removal of contractor workers, making it contractually impossible to pre-screen particular contractors based on injury/illness rates. In such cases, OSHA may accept alternative strategies to assure worker safety and health protection.

B. **Injury/Illness History.** The following additional requirements apply to Federal agency worksites:

1. Federal agency worksites must keep, in addition to their Federal Occupational Injury and Illness Log, OSHA 300 logs for the previous 3 calendar years. These will be compared to their private sector counterparts.
2. If using the calendar year creates a recordkeeping burden, Federal agency worksites may use the fiscal year to calculate injury/illness rates.

Appendix C

Appendix D
Onsite Evaluation Report Format

Verbatim copy of Appendix E, "Onsite Evaluation Report Format," OSHA Instruction Directive Number TED 8.4, *Voluntary Protection Programs (VPP): Policies and Procedures Manual*, effective date March 25, 2003.

Definitions and Acronyms

"MR" in an item means "minimum required." Regarding "minimum required" items, TED 8.4 states:

> Minimum Requirements (MR) represent those elements of a site's safety and health management system that must be in place and at least minimally effective in order for a site to be considered for participation in the Merit Program. If a site fails to meet even one MR, then it is not eligible for participation in the VPP and should be asked to withdraw its application. Requirements that are considered MR will have the symbols: **MR**.
>
> A Minimum Requirement is a basic element or sub-element that the VPP considered critical to workplace safety and health. A Minimum Requirement must be at least minimally effective for the site to avoid regular or ongoing exposure of employees to serious workplace hazards.
>
> A site is not 'minimally effective' on an aspect of VPP if the site does not have one or more of the required elements of the program in operational status or ready for implementation. As a result employees may be exposed to a serious hazard or hazards. In some cases, the site may have the element, but it so ineffective that it is as if the element did not exist.[1]

---

1. Taken from OSHA TED 8.4, Appendix F, "Site Worksheet Minimum Requirement Guidance."

Appendix D

VPP SITE REPORT
Recommending
STAR APPROVAL
For

Company Name
City, State

April 1, 2001

Report Date
April 4, 2001

Evaluation Team
Name, Team Leader
Name, Backup Team Leader
Name, Safety Specialist
Name, Hygienist
Name, SGE

I. Purpose and Scope of Review
- Site name
- Site location
- Date of evaluation
- Purpose of evaluation (Star approval, Merit approval, Demo approval)
- VPP Team Members
- VPP Volunteers

II. Methods of Data Collection
- Information on which report is based (application, previous reports, walk through, on site documentation, etc.)

III. Employees at the Worksite
- Number of employees
- Contract workers and/or temporary workers
- Collective bargaining agent(s) representing the employees
- Number of interviews conducted with different types of workers

IV. The Worksite
- SIC Code

- Site description (one location or many, acreage, age, primary structures, etc.)
- Basic description of processes, products, and applications
- Covered under Process Safety Management
- Housekeeping

V. Worksite Hazards
- Site hazards

VI. Injury and Illness Rates
- Rates—TCIR, DART
- Comparisons to BLS industry averages

VII. OSHA Activity
- Prior OSHA inspection activity
- Relationship with OSHA

VIII. Elements of the VPP Review/Program Changes
- Bulleted summary of VPP Elements with a reminder that all aspects of the Safety and Health program meet the VPP requirements as set forth in TED (Refer to the VPP Site Worksheet for specifics).
  - Management, Leadership, and Employee Involvement
  - Worksite Analysis
  - Hazard Prevention and Control
  - Safety & Health Training
    - For Reapproval evaluations, discuss significant program or site changes since the last visit. A bulleted list is acceptable.

IX. Areas of Excellence
- Bulleted list and description of best practices (e.g., machine guarding, ergonomics, lockout/tagout, employee involvement)

X. Recommendation for Participation
- Recommendation

XI. Goals
- Merit goals (if relevant)
- 1-Year Conditional goals (if relevant)

Appendix D

A review of the OSHA 200/300 logs was made. The following are the total incidence and lost workday case rates since 20XX:

Year	Hours	Total # of Cases	TCIR	Number of Cases Involving Days Away from Work, Restricted Activity or Job Transfer	DART Rate
1999	185,445	2	2.2	2	2.2
2000	216,212	2	1.9	1	0.9
2001	195,444	1	0.1	0	0.0
Total	597,791	7		5	
Three-Year Rate (1999–2001)			2.3		1.7
BLS National Average for 2001 (SIC 3728)			8.6		4.2
2002 YTD	43,315	0	0.0	0	0.0

For the period 1999-2001, the site's:

- Total Case Incidence Rate (TCIR) is X.X (XX% above/below the 2000 BLS industry averages for SIC XXXX).

- Days Away from Work, Restricted Activity or Job Transfer (DART) case incidence rate is X.X (XX% above/below the 2000 BLS industry averages for SIC XXXX).

The information on the OSHA 300 log OSHA 300 logs supports the information provided in the application, and the company's first report of injury forms support the data in the logs. The Honeywell Health, Safety and Environmental (HS&E) Coordinator is responsible for the entries to the OSHA 300 log and verified the accuracy of the records. The HS&E Coordinator understands the recordkeeping requirements. Based upon interviews conducted with management and employees, the logs accurately reflect the injury and illness experience at this plant.

There were temporary employees at the worksite at the time of the team's visit. Injuries or illnesses experienced by temporary employees under the direct supervision of Honeywell Aircraft Landing Systems are recorded on the worksite's OSHA 300 log OSHA 300 log. There was one temporary employee injury recorded on the worksite's OSHA 300 log for 2001.

VPP SITE WORKSHEET
Recommending
STAR APPROVAL
for
Company Name
City, State

Evaluation Date
April 1, 2001

Report Date
April 4, 2001

Evaluation Team
Name, Team Leader
Name, Backup Team Leader
Name, Safety Specialist
Name, Hygienist
Name, SGE

	How Assessed			
	Yes or No	Interview	Observation	Doc Review
Section 1: Management Leadership & Employee Involvement				
A. Written Safety & Health Management System				
A1. Are all the elements (such as Management Leadership and Employee Involvement, Worksite Analysis, Hazard Prevention and Control, and Safety and Health Training) and sub-elements of a basic safety and health management system part of a signed, written document? (For Federal Agencies, include 29 CFR 1960.) If not, please explain. •				
A2. Have all VPP elements and sub-elements been in place at least 1 year? If not, please identify those elements that have not been in place for at least 1 year. •				

Appendix D

A3. Is the written safety and health management system at least minimally effective to address the scope and complexity of the hazards at the site? (Smaller, less complex sites require a less complex system.) If not, please explain. **MR** •				
A4. Have any VPP documentation requirements been waived (as per FRN page 656, paragraph F5a4)? If so, please explain. •				

	How Assessed			
	Yes or No	Inter-view	Obser-vation	Doc Review
Section 1: Management Leadership & Employee Involvement				
B. Management Commitment & Leadership				
B1. Does management overall demonstrate at least minimally effective, visible leadership with respect to the safety and health program (considering FRN items F5 A-H)? Provide examples. **MR** •				
B2. How has the site communicated established policies and results-oriented goals and objectives for worker safety to employees? •				
B3. Do employees understand the goals and objectives for the safety and health program? •				
B4. Are the safety and health program goals and objectives meaningful and attainable? Provide examples supporting the meaningfulness and attainability (or lack-there-of if answer is no) of the goal(s). (Attainability can either be unrealistic/realistic goals or poor/good implementation to achieve them.) (See: TED Chapter 3 II C1a) •				

B5. How does the site measure its progress towards the safety and health program goals and objectives? Provide examples. •				

	How Assessed			
	Yes or No	Inter-view	Obser-vation	Doc Review
Section 1: Management Leadership & Employee Involvement				
C. Planning				
C1. How does the site integrate planning for safety and health with its overall management planning process (for example, budget development, resource allocation, or training)? •				
C2. Is safety and health effectively integrated into the site's overall management planning process? If not, please explain. •				

	How Assessed			
	Yes or No	Inter-view	Obser-vation	Doc Review
Section 1: Management Leadership & Employee Involvement				
D. Authority and Line Accountability				
D1. Does top management accept ultimate responsibility for safety and health in the organization? (Top management acknowledges ultimate responsibility even if some safety and health functions are delegated to others.) If not, please explain. **MR** •				

Appendix D

	Yes or No	Interview	Observation	Doc Review
D2. How is the assignment of authority and responsibility documented and communicated (for example, organization charts, job descriptions)? •				
D3. Do the individuals assigned responsibility for safety and health have the authority to ensure that hazards are corrected or necessary changes to the safety and health management system are made? If not, please explain. **MR** •				
D4. How are managers, supervisors, and employees held accountable for meeting their responsibilities for workplace safety and health? (Annual performance evaluations for managers and supervisors are required.) •				
D5. Are adequate resources (equipment, budget, or experts) dedicated to ensuring workplace safety and health? Provide examples **MR** •				
D6. Is access to experts (for example, Certified Industrial Hygienists, Certified Safety Professionals, Occupational Nurses, or Engineers), reasonably available to the site, based upon the nature, conditions, complexity, and hazards of the site? If so, under what arrangements and how often are they used? •				

	How Assessed			
	Yes or No	Interview	Observation	Doc Review
Section 1: Management Leadership & Employee Involvement				
E. Contract Workers				
E1. Does the site utilize contractors? Please explain. •				

E2. Were there contractors onsite at the time of the evaluation? •				
E3. When selecting onsite contractors, how does the site evaluate the contractor's safety and health programs and performance (including rates)? (See: TED Chapter 3 IV 3-19) •				
E4. Are contractors and subcontractors at the site to maintain effective safety and health programs and to comply with all applicable OSHA and company safety and health rules and regulations? If so, please provide examples. •				
E5. Does the site's contractor program cover the prompt correction and control of hazards in the event that the contractor fails to correct or control such hazards? Provide examples. **MR** •				
E6. How does the site document and communicate oversight, coordination, and enforcement of safety and health expectations to contractors? •				
E7. Have the contract provisions specifying penalties for safety and health issues been enforced, when appropriate? If not, please explain. •				
E8. How does the site monitor the quality of the safety and health protection of its contract employees? •				
E9. If the contractors' injury and illness rates are above the average for their industries, does the site have procedures that ensure all employees are provided effective protection on the worksite? If not, please explain. •				

Appendix D

E10. Do contract provisions for contractors require the periodic review and analysis of injury and illness data? Provide examples. •				
E11 Based on your answers to the above items, is the contract oversight minimally effective for the nature of the site? (Inadequate oversight is indicated by significant hazards created by the contractor, employees exposed to hazards, or a lack of host audits.) If not, please explain. **MR** •				

	How Assessed			
	Yes or No	Inter-view	Obser-vation	Doc Review
Section 1: Management Leadership & Employee Involvement				
F. Employee Involvement				
F1. How were employees selected to be interviewed by the VPP team? •				
F2. How many employees were interviewed formally? How many were interviewed informally? •				
F3. Do employees support the site's participation in the VPP Process? **MR** •				
F4. Do employees feel free to participate in the safety and health management system without fear of discrimination or reprisal? If so, please explain. **MR** •				

F5. Please describe at least three ways in which employees are meaningfully involved in the problem identification and resolution, or evaluation of the safety and health program (beyond hazard reporting). (See: FRN Chapter 3 Paragraph II.C.1.b) •				
F6. Are employees knowledgeable about the site's safety and health management system? If not, please explain. •				
F7. Are employees knowledgeable about the VPP program? If not, please explain. •				
F8. Are the employees knowledgeable about OSHA rights and responsibilities? If not, please explain. •				
F9. Do employees have access to results of self-inspection, accident investigation, appropriate medical records, and personal sampling data upon request? If not, please explain. •				

Section I: Management Leadership & Employee Involvement
Merit Goals: Include cross reference to section, subsection, and questions, e.g. I.B2.
1. 2.

90-Day Items: (Delete this section for final transmittal to National Office)
1. 2.

Appendix D

Best Practices:
1.
2.

Comments Including Recommendations: (optional)
1.
2.

Documents Referenced, Programs Reviewed: (optional)
1.
2.

	How Assessed			
	Yes or No	Interview	Observation	Doc Review
Section II: Worksite Analysis				
A. Baseline Hazard Analysis				
A1. Has the site been at least minimally effective at identifying and documenting the common safety and health hazards associated with the site (such as those found in OSHA regulations, building standards, etc., and for which existing controls are well known)? If not, please explain. **MR** •				
A2. What methods are used in the baseline hazard analysis to identify health hazards? (Please include examples of instances when initial screening and full-shift sampling were used. See FRN page 45657, F5.B.2.b.) •				

238

A3. Does the site have a documented sampling strategy used to identify health hazards and assess employees' exposure (including duration, route, and frequency of exposure), and the number of exposed employees? If not, please explain. •				
A4. Do sampling, testing, and analysis follow nationally recognized procedures? If not, please explain. •				
A5. Does the site compare sampling results to the minimum exposure limits or are more restrictive exposure limits (PELs, TLVs, etc.) used? Please explain. •				
A6. Does the baseline hazard analysis adequately identify hazards (including health) that need further analysis? If not, please explain. •				
A7. Do industrial hygiene sampling data, such as initial screening or full shift sampling data, indicate that records are being kept in logical order and include all sampling information (for example, sampling time, date, employee, job title, concentrated measures, and calculations)? If not, please explain the deficiencies and how they are being addressed. •				

Appendix D

	How Assessed			
	Yes or No	Inter-view	Obser-vation	Doc Review
Section II: Worksite Analysis				
B. Hazard Analysis of Significant Changes				
B1. When purchasing new materials or equipment, or implementing new processes, what types of analyses are performed to determine their impact on safety and health? Is it adequate? •				
B2. When implementing/introducing non-routine tasks, materials or equipment, or modifying processes, what types of analyses are performed to determine their impact on safety and health? Is it adequate? •				

	How Assessed			
	Yes or No	Inter-view	Obser-vation	Doc Review
Section II: Worksite Analysis				
C. Hazard Analysis of Routine Activities				
C1. Is there at least a minimally effective hazard analysis system in place for routine operations and activities? **MR** •				
C2. Does hazard identification and analysis address both safety and health hazards, if appropriate? If not, please explain. •				
C3. What hazard analysis technique(s) are employed for routine operations and activities (e.g., job hazard analysis, HAZ-OPS, fault trees)? Are they adequate? •				

		How Assessed		
	Yes or No	Inter-view	Obser-vation	Doc Review
C4. Are the results of the hazard analysis of routine activities adequately documented? If not, please explain. •				

	How Assessed			
	Yes or No	Inter-view	Obser-vation	Doc Review
Section II: Worksite Analysis				
D. Routine Inspections				
D1. Does the site have a minimally effective system for performing safety and health inspections (i.e., a minimally effective system identifies hazards associated with normal operations)? If not, please explain. **MR** •				
D2. Are routine safety and health inspections conducted monthly, with the entire site covered at least quarterly (for construction: entire site weekly)? •				
D3. How do inspections use information discovered through the baseline hazards analysis, job hazard analysis, accident/incident analysis, employee concerns, sampling results, etc.? •				
D4. Are those personnel conducting inspections adequately trained in hazard identification? If not, please explain. •				
D5. Is the routine inspection system written, including documentation of results? If not, please explain. •				
D6. Do the written routine inspection reports clearly indicate what needs to be corrected, by whom, and by when? If not, please explain. •				

Appendix D

D7. Did the VPP team find hazards that should have been found through self-inspection? If so, please explain. •				

Section II: Worksite Analysis

	How Assessed			
	Yes or No	Inter-view	Obser-vation	Doc Review
E. Hazard Reporting				
E1. Does the site have a reliable system for employees to notify appropriate management personnel in writing about safety and health concerns? Please describe. •				
E2. Do the employees agree that they have an effective system for reporting safety and health concerns? If not, please explain. •				
E3. Is there a minimally effective means for employees to report hazards and have them addressed? If not, please explain. **MR** •				

Section II: Worksite Analysis

	How Assessed			
	Yes or No	Inter-view	Obser-vation	Doc Review
F. Hazard Tracking				
F1. Does the hazard tracking system address hazards found by employees, hazard analysis of routine and non-routine activities, inspections, and accident or incident investigations? If not, please explain. •				

		How Assessed		
	Yes or No	Interview	Observation	Doc Review
F2. Does the tracking system result in hazards being corrected and provide feedback to employees for hazards they have reported. If not, please explain. •				
F3. Does the tracking system result in timely correction of hazards with interim protection established when needed? Please describe. If not, please explain. •				
F4. Does a minimally effective tracking system exist that results in hazards being controlled? If not, please explain. **MR** •				

	How Assessed			
	Yes or No	Interview	Observation	Doc Review
Section II: Worksite Analysis				
G. Accident/Incident Investigations				
G1. Is there a minimally effective system for conducting accident/incident investigations, including near-misses? If not, please explain. **MR** •				
G2. Are those conducting the investigations trained in accident/incident investigation techniques? If not, please explain. •				
G3. Describe how investigations discover and document all the contributing factors that led to an accident/incident. •				

Appendix D

G4. Were any hazards discovered during the investigation previously addressed in any prior hazard analyses (e.g., baseline, self-inspection)? If not, please explain. •				

	How Assessed			
	Yes or No	Interview	Observation	Doc Review
Section II: Worksite Analysis				
H. Safety and Health Program Evaluation				
H1. Briefly describe the system in place for conducting an annual evaluation. •				
H2. Does the annual evaluation cover the aspects of the safety and health program, including the elements described in the **Federal Register**? If not, please explain. •				
H3. Does the annual evaluation include written recommendations in a narrative format? If not, please explain. •				
H4. Is the annual evaluation an effective tool for assessing the success of the site's safety and health system? Please explain. •				
H5. What evidence demonstrates that the site responded adequately to the recommendations made in the annual evaluation? •				

	How Assessed			
	Yes or No	Inter-view	Obser-vation	Doc Review
Section II: Worksite Analysis				
I. Trend Analysis				
I1. Does the site have a minimally effective means for identifying and assessing trends? **MR** •				
I2. Have there been any injury and/or illness trends over the last three years? If so, please explain. •				
I3. If there have been injury and/or illness trends, what courses of action have been taken? Are they adequate? •				
I4. Does the site assess trends utilizing data from hazard reports or accident/incident investigations to determine the potential for injuries and illnesses? If not, please explain. •				

Section II: Worksite Analysis

Merit Goals: Include cross reference to section, subsection, and questions, e.g. I.B2.

1.

2.

90-Day Items: (Delete this section for final transmittal to National Office)

1.

2.

Appendix D

Best Practices:
1.
2.

Comments Including Recommendations: (optional)
1.
2.

Documents Referenced, Programs Reviewed: (optional)
1.
2.

	How Assessed			
	Yes or No	Interview	Observation	Doc Review
Section III: Hazard Prevention and Control				
A. Hazard Prevention and Control				
A1. Does the site select at least minimally effective controls to prevent exposing employees to hazards. **MR** •				
A2. When the site selects hazard controls, does it follow the preferred hierarchy (engineering controls, administrative controls, work practice controls [e.g lockout/tag out, bloodborne pathogens, and confined space programs], and personal protective equipment) to eliminate or control hazards? Please provide examples, such as how exposure to health hazards were controlled. •				

A3. Describe any administrative controls used at the site to limit employee exposure to hazards (for example, job rotation). •				
A4. Do the work practice controls and administrative controls adequately address those hazards not covered by engineering or administrative controls? If not, please explain. •				
A5. Are the work practice controls (e.g. lockout/tag out, blood born pathogens, and confined space programs) recommended by hazard analyses implemented at the site? If not, please explain. •				
A6. Are follow-up studies (where appropriate) conducted to ensure that hazard controls were adequate? If not, please explain. •				
A7. Are hazard controls documented and addressed in appropriate procedures, safety and health rules, inspections, training, etc.? Provide examples. •				
A8. Are there written worker safety procedures including a disciplinary system? Describe the disciplinary system. •				
A9. Has the disciplinary system been enforced equally for both management and employees, when appropriate? If not, please explain. •				
A10. Does the site have minimally effective written procedures for emergencies (TED 3-16 3h)? **MR** •				
A11. Are emergency drills held at least annually? •				

Appendix D

A12. Does the site have a written preventative/predictive maintenance system? If not, please explain. •				
A13. Did the hazard identification and analysis (including manufacturers' recommendations) identify hazards that could result if equipment is not maintained properly? If not, please explain. •				
A14. Does the preventive maintenance system adequately detect hazardous failures before they occur? If not, please explain. •				
A15. How does the site select Personal Protective Equipment (PPE)? •				
A16. Do employees understand the limitations and uses of PPE? If not, please explain. •				
A17. Did the team observe employees using, storing, and maintaining PPE properly? If not, please explain. •				
A18. Is the site covered by the Process Safety Management Standard (29 CFR 1910.119)? If not, skip to section B. •				
A19. Which chemicals that trigger the Process Safety Management (PSM) standard are present? •				
A20. Please describe the PSM elements in place at the site (do not duplicate if included elsewhere in the report, such as under contractors, preventive maintenance, emergency response, or hazard analysis). •				

	How Assessed			
	Yes or No	Interview	Observation	Doc Review
Section III: Hazard Prevention and Control				
B. Occupational Health Care Program and Recordkeeping				
B1. Describe the occupational health care program (including availability of physician services, first aid, and CPR/AED) and special programs such as audiograms or other medical tests used. •				
B2. How are licensed occupational health professionals used in the site's hazard identification and analysis, early recognition and treatment of illness and injury, and the system for limiting the severity of harm that might result from workplace illness or injury? Is this use appropriate? •				
B3. Is the occupational health program adequate for the size and location of the site, as well as the nature of hazards found here? If not, please explain. •				

Section III: Hazard Prevention and Control

Merit Goals: Include cross reference to section, subsection, and questions, e.g. I.B2.

1.

2.

90-Day Items: (Delete this section for final transmittal to National Office)

1.

2.

Appendix D

Best Practices:
1.
2.

Comments Including Recommendations: (optional)
1.
2.

Documents Referenced, Programs Reviewed: (optional)
1.
2.

	How Assessed			
	Yes or No	Inter-view	Obser-vation	Doc Review
Section IV: Safety and Health Training				
A. Safety and Health Training				
A1. What are the safety and health training requirements for managers, supervisors, employees, and contractors? •				
A2. Who delivers the training? •				
A3. How are the safety and health training needs for employees determined? •				

A4. Does the site provide minimally effective training to educate employees regarding the known hazards of the site and their controls? If not, please explain. **MR** •				
A5. What system is in place to ensure that all employees and contractors have received and understand the appropriate training? •				
A6. Who is trained in hazard identification and analysis? •				
A7. Is training in hazard identification and analysis adequate for the conditions and hazards of the site? If not, please explain. •				
A8. Does management have a thorough understanding of the hazards of the site? Provide examples that demonstrate their understanding. •				

Section III: Hazard Prevention and Control

Merit Goals: Include cross reference to section, subsection, and questions, e.g. I.B2.

1.

2.

90-Day Items: (Delete this section for final transmittal to National Office)

1.

2.

Appendix D

Best Practices:

1.
2.

Comments Including Recommendations: (optional)

1.
2.

Documents Referenced, Programs Reviewed: (optional)

1.
2.

VPP TEAM COMPOSITION AND PROGRAM ACTIVITIES DATA SHEET

Site Name:

Site Name and Address:

Small Business? (< 250 employees onsite and < 500 employees corporate-wide) (Y/N):
Region:
Dates of Onsite:
Date of 90-day Item Completion (if applicable):

Check one: ___ Approval ___ Re-Approval

Report drafted onsite: ___ Yes ___ No

Team Composition			
Last Name	ID*	Role	OSHA Volunteer

* The ID is the first letter of the last name, plus the last four digits of the person's social security #.

BEST PRACTICES CHECKLIST

___ ERGO Program   ___ Confined Space   ___ Lockout/Tagout

___ PSM   ___ Hazard Analysis   ___ Contractor

___ Inspections   ___ Accountability   ___ Tracking

___ Employee Involvement   ___ Industrial Hygiene

___ Emergency Response Teams   ___ English/Second Lang.

___ Other (describe)

Appendix D

STRATEGIC PLAN FOCUS AREA

____ Amputations   ____ Lead Exposures   ____ Nursing Homes

____ Silica   ____ Ergonomics   ____ Logging

____ Food Processing   ____ Shipyards   ____ Construction

Appendix E
Recommended Interview Questions

Verbatim copy of OSHA Instruction Directive Number TED 8.4, *Voluntary Protection Programs (VPP): Policies and Procedures Manual,* Appendix G, "Recommended Interview Questions," effective date: March 25, 2003

I. **Purpose.** Interviews are an important tool in assessing the effectiveness of a site's safety and health programming. These questions are intended to guide the OSHA reviewer during oral employee interviews. To begin, explain the purpose of the interview and the reason for OSHA's presence at the site. Make employees aware that interviews are kept confidential and that the employee's responses will not in themselves determine company approval or disapproval.

II. **General Employee Interview Questions**

   A. How long have you worked here?

   B. Tell me about your job. What do you do during a typical day?

   C. What are the safety and health hazards of your job?

   D. How do you protect yourself from those hazards? What kind of personal protective equipment do you wear? Were you provided training?

   E. What type of safety and health training have you received?

   F. What happens if management disobeys a company safety rule? If an employee disobeys?

   G. How do you respond in the event of a fire, hazardous waste spill, alarm, or medical emergency?

   H. What does VPP mean to you?

   I. What is one method of reporting a safety or health concern? What was the last unsafe practice you reported and/or corrected?

   J. How do your supervisors demonstrate their involvement in safety and health?

Appendix E

K. Have you ever seen anyone testing the air, noise levels, or conducting other surveys for possible health hazards? Do you know what the results were or what they meant?

L. Have you or anyone you know ever been injured or experienced a job related illness? What is the procedure when someone is injured?

M. How are you involved in the safety decision-making process?

N. Is safety and health valued in your organization?

O. What is one objective in your department's safety program?

P. How does management support your involvement in safety?

Q. What are your rights under OSHA?

R. Is there anything else you think we should know about the safety and health program here?

III. **Supervisors**

A. How long have you worked here? When did you become a supervisor?

B. What do you see as your role in safety and health?

C. To what kinds of hazards are you and/or your employees exposed?

D. Has the company's upper management provided adequate resources for safety and health programming, such as funding, time, and technical support?

E. What do you do when you discover a hazard in your area?

F. What do you do when an employee reports a hazard in your area?

G. Do you provide employee training in safety related topics? (If so, please describe.)

H. Please give some examples where you had to use the disciplinary system for infractions of safety and health rules.

I. When was the last emergency drill? What is your role in drills?

J. How are you held accountable for ensuring safe and healthful working conditions in your area?

K. At high hazard chemical plants only: Is maintenance satisfactory, particularly on release prevention equipment? Is there adequate supervision provided for work performed on all shifts?

L. Do you have contract employees working in your area? If so, how do you control and address safety or health hazards relating to or created by them?

M. Are there routine or unannounced inspections? Who participates?

IV. **Administrators and Executives**
   A. How long have you been with (company)?
   B. Describe the type of safety and health hazards at this site.
   C. How does management ensure that employee exposure to those hazards is eliminated or controlled?
   D. How do you demonstrate leadership in and commitment to safety and health?
   E. What benefits will a VPP partnership provide for your company?
   F. What do you think are your facility's "best practices" in safety and health?
   G. How do you address the competing pressures of production and safety?
   H. How do you hold your supervisors accountable for safety and health? Have you ever had to discipline a supervisor for not following the rules?
   I. How are you held accountable for your safety and health responsibilities?

V. **Recordkeepers**
   A. Who is responsible for recordkeeping?
   B. Is your site recordkeeping centralized? Is it computerized?
   C. Do you have a completed Summary of Occupational Injuries and Illnesses for the last 3 calendar years? Do you have the supplemental documentation for each case entered on the log?
   D. Which form do you use as the supplementary record: OSHA's First Report of Injury, a State workers' compensation form, an insurer's form, or other?
   E. What is the process by which injury and illness information gets to the recordkeeper? After an injury or illness occurs, how long does it take to enter it on the log?
   F. What type of reference material do you refer to for guidance on keeping illness and injury records?
   G. Who decides whether or not a case is recordable?
   H. How do you determine whether or not a case is work related?
   I. Do you record any cases on the OSHA forms that are not com-

Appendix E

        pensable under workers' compensation?

    J. How do you distinguish between an injury and an illness? Between medical treatment and first aid?

    K. When does a case involve lost workdays? What constitutes restricted work activity?

    L. What is your process for monitoring applicable contractor logs?

    M. How do you safeguard the confidentiality of medical records?

    N. How do you assure that any work restrictions are applied appropriately?

    O. How have you assured timely and clear communications with the health care professional?

VI. **Occupational Health Care Professionals**

    A. What are your qualifications and licenses?

    B. What procedures are in place to ensure that health care services are delivered consistently and effectively?

    C. What type of audit procedures do you use to compare your process with acceptable standards of practice and OSHA requirements?

    D. Are employees provided timely access to services?

    E. How do you assure that work restrictions or work removal are followed?

    F. How are you made aware of the job hazards at this facility? Are you included in identification of workplace hazards, or development of restricted duty jobs, or other onsite issues?

    G. What kinds of health surveillance programs are in place?

    H. How do you communicate health surveillance data to employees and management to reduce future risk?

    I. Explain how you evaluate the effectiveness of your occupational health care program.

VII. **Maintenance Personnel**

    A. Is there a scheduled preventive maintenance program? How is it carried out?

    B. At sites covered by Process Safety Management (PSM): Does the preventive maintenance program include:

        1. Critical instrumentation and controls?

2. Pressure relief devices and systems?
3. Metals inspection?
4. Environmental controls, scrubbers, filters, etc.?

C. At PSM sites: Does the design, inspection, and maintenance activity include procedures to prevent piping cross-connections between potable water systems and non-potable systems?

1. How are these procedures carried out?
2. How are systems monitored and inspected to find any cross-connections?

D. Do maintenance personnel participate in safety functions?

E. Is there a priority system for safety/environmental related maintenance items? Is it being followed?

F. Does the preventive maintenance program include onsite vehicles, sprinkler systems, detection/alarm equipment, fire protection and emergency equipment?

G. Do you have input concerning safety and ease of maintenance for new equipment and machinery purchases?

H. Do you have an inventory of spare parts critical to safety and environmental protection?

I. Are you trained in the control of hazardous energy and the proper use of locks and tags?

J. Is there a system in place to track requests for repairs?

K. What methods are used to monitor the condition of critical equipment?

L. What is the ratio of scheduled versus unscheduled maintenance work?

M. What has the trend been like over the past few years?

VIII. General Questions for Onsite Evaluations to Determine Reapproval

A. Describe any changes in your job or in the handling of safety issues since the last OSHA onsite evaluation.

B. How familiar are you with VPP? Has your awareness increased since the last visit?

C. Do you have any increased knowledge of your rights under the program, including your right to receive upon request results of

Appendix E

   self-inspections or accident investigations?
D. Do you feel that the VPP partnership has had a positive impact on your job and your safety?
E. Have you noticed any changes in safety and health conditions here since the site's approval in VPP?

Appendix F
Environmental Compliance Information

Environmental Protection Agency (EPA)

The EPA's website (www.epa.gov) provides general information about the EPA and links to other websites and EPA information.

For environmental publications, subscriptions, and general information contact the EPA National Service Center for Environmental Publications (NSCCP).

US EPA/NSCCP
PO Box 42419
Cincinnati OH 45242-0419
Phone: (800) 490-9298 or (513) 489-8190
Fax: (513) 489 8695 (24 hours day/7 days a week)
Web: www.epa.gov/ncepihom

Internal Organization of Standards (ISO)

The ISO develops the environmental standards for the world. You can find a vast storehouse of information on the ISO on the internet—the ISO website (www.iso.ch/iso/en/ISOOnline.openerpage) is a good place to start.

As published on the EPA's Enviroene website (www.epa.gov/envirosense, ISO is:

> ...a private sector, international standards body based in Geneva, Switzerland. Founded in 1947, ISO promotes the international harmonization and development of manufacturing, product and communications standards. ISO has promulgated more than 8,000 internationally accepted standards for everything from paper sizes to film speeds. More than 120 countries belong to ISO as full voting members, while several other countries serve as observer members. The United States is a full voting member and is officially represented by the American National Standards Institute (ANSI).
>
> ISO produces internationally harmonized standards through a structure of Technical Committees (TCs). The TCs usually

divide into Subcommittees which are further subdivided in Working Groups where the actual standards writing occurs."

ISO 14000 Standard for Environmental Management

The ISO 14000 Information Center website (www.iso14000.com) provides information on this guiding document for environmental management systems that meets EPA requirements.

EPA has endorsed the use of environmental management systems (EMSs) as contained in ISO 14000.

Other Sources of Information

Another source to access is Standards Council of Canada website (www.scc.ca) for extensive general information.

Some large American companies post their environmental programs on their websites, along with their other safety and health activities. One example is ExxonMobil's website (www.exxonmobil.com). Their Publications page contains the company's guiding principles, SH&E policies, and annual SH&E report.

Another example is BP (formerly BP Amoco). BP's website (www.bp.com) yielded several topics. Keyword searches for "environmental program" and "environment" return many pages of information.

Other sources of information are the magazines and publications about EPA requirements, news about environmental products, activities, and the current state of our environment. A keyword search for "environmental magazines" returns hundreds of hits.

Books on environmental issues are also found at any of the online bookstores, such as amazon.com or bn.com.

Appendix G
Program Evaluation Profile (PEP) Example

CAUTION: The following excerpts are from the OSHA Notice CPL 2 distributed by the Directorate of Compliance Programs on August 1, 1996. This Program Evaluation Profile (PEP) notice was *cancelled on November 15, 1996*. This document is provided as an **example** of an auditing tool to assist in the evaluation of a safety and health program.

The Program Evaluation Profile (PEP)

Purpose

This notice establishes policies and procedures for the Program Evaluation Profile (PEP), Form OSHA-195, which is to be used in assessing employer safety and health programs in general industry workplaces.

References

1. OSHA Instruction CPL 2.103, September 26, 1994, Field Inspection Reference Manual (FIRM).
2. OSHA Instruction CPL 2.111, November 27, 1995, Citation Policy for Paperwork and Written Program Requirement Violations.
3. Safety and Health Program Management Guidelines, Federal Register, January 26, 1989, Vol. 54, No. 16, pp. 3904-3916.

Background

OSHA's assessment of safety and health conditions in the workplace depends on a clear understanding of the programs and management systems that an employer is using for safety and health compliance. The Agency places a high priority on safety and health programs and wishes to encourage their implementation.

1. **Evaluation of Workplace Safety and Health Programs.** In the past, compliance officers have evaluated employers' safety and health programs, but those evaluations have not always required thorough documentation in case files. More detailed evaluation and documentation is now required to meet the Agency's need to assess such programs

Appendix G

accurately and to respond to workplace compliance conditions accordingly.

2. **Safety and Health Program Management Guidelines.** In January 1989, OSHA published its voluntary Safety and Health Program Management Guidelines (*Federal Register*, Vol. 54, No. 16, pp. 3904-3916; hereinafter referred to as the 1989 Guidelines), which have been widely used in assessing employer safety and health programs.

3. **The PEP.** Appendix A of this directive contains the Program Evaluation Profile (PEP), Form OSHA-195, a new program assessment instrument.

    a. The PEP was developed by representatives of OSHA's National Office and field staff in a cooperative effort with the National Council of Field Labor Locals (NCFLL).

    b. The PEP is presented in a format that will enable the compliance officer to present information about the employer's program graphically. While the PEP is compatible with other evaluation tools based on the 1989 Guidelines, it is not the only such tool that will be used. It is not a substitute for any other kinds of program evaluations conducted by OSHA, such as those that are required by OSHA standards. Program evaluations, such as those required by the Process Safety Management Standard, the Lockout/Tagout standard, the Bloodborne Pathogens standard, and others, are considered to be an integral part of a good safety and health program if the workplace is covered by those standards, and therefore must be included in the PEP review.

    c. Instructions for use of the PEP are found in 'Using the PEP' section of this notice.

Application

1. The PEP shall be completed for general industry inspections and compliance-related activities that include an evaluation of an employer's workplace safety and health programs.

    a. The PEP is an educational document for workers and employers, as well as a source of information for OSHA's use in the inspection process.

    b. A new PEP need not be completed when a PEP has recently been done for a specific workplace and the compliance officer judges that no substantive changes in the employer's safety and health program have occurred.

    c. In multi-employer workplaces, a PEP shall be completed for the safety and health program of the host employer. This PEP will normally apply to all subordinate employers onsite, and individual PEPs need not be completed for them. The compliance officer may, however, complete a PEP for any other employers onsite for which he/she believes it is appropriate; e.g., where a subcontractor's program is markedly better or worse than that of the host employer.

    d. The PEP shall be used in experimental programs and cooperative compliance programs (e.g., Maine 200) that require evaluation of an employer's safety and health program, except where other program evaluation methods/tools are specifically approved.

2. The compliance officer's evaluation of the safety and health program contained in the PEP shall be shared with the employer and with employee representatives, if any, no later than the date of issuance of citations (if any).

    a. The preliminary assessment from the PEP shall be discussed with the employer in the closing conference; the employer shall be advised that this assessment may be modified based on further inspection results or additional information supplied by the employer.

    b. The compliance officer's evaluation of the safety and health program may be shared with the employer and with employee representatives in the following ways:

        (1). Giving a copy of the completed PEP to the employer and employee representatives.

        (2). Giving a blank PEP to the employer and employee representatives, who may complete it themselves based on compliance officer comments and their own knowledge.

        (3). Providing only verbal comments and recommendations to the employer and employee representatives.

        (4). Providing written comments and recommendations (e.g., in a letter from the Area Director) to the employer and employee representatives.

    c. Because the PEP represents the compliance officer's evaluation of an employer's worksite safety and health program at the time an inspection was conducted, the scoring shall normally be modified only by the inspecting compliance officer, except in the case of clear errors (e.g., computation).

Appendix G

Using the PEP

The PEP will be used as a source of safety and health program evaluation for the employer, employees, and OSHA.

1. **Gathering Information for the PEP** begins during the opening conference and continues through the inspection process.

   a. The compliance officer shall explain the purpose of the PEP and obtain information about the employer's safety and health program in order to make an initial assessment about the program.

   b. This initial assessment shall be verified--or modified--based on information obtained in interviews of an appropriately representative number of employees and by observation of actual safety and health conditions during the inspection process.

   c. If the employer does not wish to volunteer the information needed for the PEP, the compliance officer shall note this in the case file but shall not press the issue. The benefits of a PEP evaluation shall, however, be explained.

2. **Recording the Score.** The program elements in the PEP correspond generally to the major elements of the 1989 Guidelines.

   a. **Elements.** The **six** elements to be scored in the PEP are:

   (1). Management Leadership and Employee Participation.

   (2). Workplace Analysis.

   (3). Accident and Record Analysis.

   (4). Hazard Prevention and Control.

   (5). Emergency Response.

   (6). Safety and Health Training.

   b. **Factors.** These elements [except for (6), Training] are divided into factors, which will also be scored. The score for an element will be determined by the factor scores. The factors are:

   (1). Management Leadership and Employee Participation.
   - Management leadership.
   - Employee participation.
   - Implementation (tools provided by management, including budget, information, personnel, assigned responsibility, adequate expertise and authority, line accountability, and program review procedures).

- Contractor safety.

(2). Workplace Analysis.
- Survey and hazard analysis.
- Inspection.
- Reporting.

(3). Accident and Record Analysis.
- Investigation of accidents and near-miss incidents.
- Data analysis.

(4). Hazard Prevention and Control.
- Hazard control.
- Maintenance.
- Medical program.

(5). Emergency Response.
- Emergency preparedness.
- First aid.

(6). Safety and Health Training (as a whole).

c. **Scoring**. The compliance officer shall objectively score the establishment on each of the individual factors and elements after obtaining the necessary information to do so. (See "Using the PEP" 5., below.) These shall be given a score of 1, 2, 3, 4, or 5.

(1). Appendix B of this notice contains the **PEP Tables**, which provide verbal descriptors of workplace characteristics for each factor for each of the five levels. Compliance officers shall refer to these tables as appropriate to ensure that the score they assign to a factor corresponds to the descriptor that best fits the worksite.

NOTE: The descriptors are intended as brief illustrations of a workplace at a particular level. In exercising their professional judgement, compliance officers should proceed with the understanding that the descriptor that "best fits" will not necessarily match the workplace exactly or in literal detail.

(2). Determine scores for each of the six elements as follows:

(a). The score for the "Management Leadership and Employee Participation" element shall be whichever is the lowest of the following:

Appendix G

1 The score for the "Management Leadership" factor.

2 The score for the "Employee Participation" factor.

3 The average score for all four factors.

NOTE: The factors of "Management Leadership" and "Employee Participation" are given greater weight because they are considered the foundation of a safety and health program.

(b). For the sixth element, *Training*, just determine the level 1-5 that best fits the worksite and note it in the appropriate box on the PEP.

(c). For each of the other four elements, **average** the scores for the factors.

(d). In **averaging** factor scores, round to the nearest whole number (1, 2, 3, 4, or 5). Round up from one-half (.5) or greater; round down from less than one-half (.5).

(3). If the employer declines to provide pertinent information regarding one or more factors or elements, a score of 1 shall be recorded for the factor or element.

(4). If the element or factor does not apply to the worksite being inspected, a notation of **"Not Applicable"** shall be made in the space provided. This shall be represented by **"N/A"** or, in IMIS applications, "**0.**" This shall not affect the score.

d. **Overall Score.** An "Overall Score" for the worksite will be recorded on the score summary. This will be the **average** of the six individual scores for elements, rounded to the nearest whole number (1, 2, 3, 4, or 5). Round up from one-half (.5) or greater; round down from less than one-half (.5).

EXAMPLE: A PEP's element scores are:

2 + 2 + 1 + 3 + 2 + 3 = 13

13 ÷ 6 = 2.16 = 2

e. **Rating and Tracking.** The six individual element scores, in sequence (e.g., "2-2-1-2-3-1") will constitute a "rating" for purposes of tracking improvements in an establishment's safety and health program, and shall be recorded.

3. **Program Levels.** The Overall Score on the PEP constitutes the "level" at which the establishment's safety and health program is scored. **Remember: this level is a relatively informal assessment of the program, and it does**

not represent a *compliance* judgement by OSHA—that is, it does not determine whether an employer is in compliance with OSHA standards. The following chart summarizes the levels:

Score	Level of Safety and Health Program
5	Outstanding program
4	Superior program
3	Basic program
2	Developmental program
1	No program or ineffective program

4. **Specific Scoring Guidance.** The following shall be taken into account in assessing specific factors:

   a. **Written Programs.** Employer safety and health programs should be in writing in order to be effectively implemented and communicated.

   (1). Nevertheless, a program's **effectiveness** is more important than whether it is in writing. A small worksite may well have an effective program that is not written, but which is well understood and followed by employees.

   (2). In assessing the effectiveness of a safety and health program that is not in written form, compliance officers should follow the general principles laid out in OSHA Instruction CPL 2.111, "Citation Policy for Paperwork and Written Program Requirement Violations." That is:

   (a). An employer's failure to comply with a paperwork requirement is normally penalized only when there is a serious hazard related to this requirement.

   (b). An employer's failure to comply with a written program requirement is normally not penalized if the employer is actually taking the actions that are the subject of the requirement.

   (3). Thus, compliance officers should follow the general principle that "performance counts more than paperwork." Neither the 1989 Guidelines nor the PEP is a standard; neither can be enforced through the issuance of citations. In using the PEP, the compliance officer is responsible for evaluating the employer's actual management of safety and health in the workplace, not just the employer's documentation of a safety and health program.

b. **Employee Participation.**

(1). Employee involvement in an establishment's safety and health program is essential to its effectiveness. Thus, evaluation of safety and health programs must include objective assessment of the ways in which workers' rights under the OSH Act are addressed in form and practice. The PEP Tables include helpful information in this regard.

(2). Employee involvement should also include participation in the OSHA enforcement process; e.g., walkaround inspections, interviews, informal conferences, and formal settlement discussions, as may be appropriate. Many methods of employee involvement may be encountered in individual workplaces.

c. **Comprehensiveness.** The importance of a safety and health program's comprehensiveness is implicitly addressed in Workplace Analysis under both "Survey and hazard analysis" and "Data analysis." An effective safety and health program shall address all known and potential sources of workplace injuries and illnesses, whether or not they are covered by a specific OSHA standard. For example, lifting hazards and workplace violence problems should be addressed if they pertain to the specific conditions in the establishment.

d. **Consistency with Violations/Hazards Found.** The PEP evaluation and the scores assigned to the individual elements and factors should be consistent with the types and numbers of violations or hazards found during the inspection and with any citations issued in the case. As a general rule, high scores will be inconsistent with numerous or grave violations or a high injury/illness rate. The following are examples for general guidance:

(1). If applicable OSHA standards require training, but the employer does not provide it, the PEP score for "Training" should not normally exceed "2."

(2). If hazard analyses (e.g., for permit-required confined spaces or process safety management) are required but not performed by the employer, the PEP score for "Workplace analysis" should not normally exceed "2."

(3). If the inspection finds numerous serious violations--in particular, high-gravity serious violations--relative to the size and type of workplace, the PEP score for "Hazard prevention and Control" should not normally exceed "2."

5. **Scope of the PEP Review.** The duration of the PEP review will vary depending on the circumstances of the workplace and the inspection. In all cases, however, this review shall include:

   a. A review of any appropriate employer documentation relating to the safety and health program.

   b. A walkaround inspection of pertinent areas of the workplace.

   c. Interviews with an appropriate number of employer and employee representatives.

Appendix A
The Program Evaluation Profile (PEP)

6. Each of the elements and factors of the PEP may be scored from 1 to 5, indicating the level of the safety and health program, as follows:

Score	Level of Safety and Health Program
5	Outstanding program
4	Superior program
3	Basic program
2	Developmental program
1	No program or ineffective program

Scoring. Score the establishment on each of the factors and elements after obtaining the necessary information to do so. These shall be given a score of 1, 2, 3, 4, or 5.

- Refer to the **PEP Tables**, Appendix B of this notice, as appropriate, to ensure that the score given to a factor corresponds to the descriptor that best fits the worksite. Determine scores for each of the six elements as follows:

- The score for the *Management Leadership and Employee Participation* element shall be whichever is the lowest of the following:

- The score for the "Management Leadership" factor, or

- The score for the "Employee Participation" factor, or

- The average score for all four factors.

- For the sixth element, *Training*, just determine the level 1-5 that best fits the worksite and note it in the appropriate box on the PEP.
- For each of the other four elements, **average** the scores for the factors.
- If the employer declines to provide pertinent information regarding one or more factors or elements, a score of 1 shall be recorded for the factor or element.
- If the element or factor does not apply to the worksite being inspected, a notation of **"Not Applicable"** shall be made in the space provided. This shall be represented by "N/A" or, in IMIS applications, "0." This shall not affect the score.

Overall Score. An "Overall Score" for the worksite will be recorded on the PEP. This will be the **average** of the six individual scores for elements, rounded to the nearest whole number (1, 2, 3, 4, or 5). Round up from one-half (.5) or greater; round down from less than one-half (.5).

EXAMPLE: A PEP's element scores are:

$2.5 + 2.7 + 2.3 + 3.0 + 2.3 + 2.0 = 14.8$

$14.8 \div 6 = 2.467 = 2$ PEP score

Appendix B

The PEP Tables

- The text in each block provides a description of the program element or factor that corresponds to the level of program that the employer has implemented in the workplace.
- To avoid duplicative language, each level should be understood as containing all positive factors included in the level below it. Similarly, each element score should be understood as containing all positive factors of the element scores below it. That is, a 3 is at least as good as a 2; a 4 is at least as good as a 3, and so on.
- The descriptors are intended as brief illustrations of a workplace at a particular level. In exercising their professional judgement, compliance officers should proceed with the understanding that the descriptor that "best fits" will not necessarily match the workplace exactly or in literal detail.

PEP
Program Evaluation Profile

Employer:
Inspection No.
Date:
CSHO ID:

			5	4	3	2	1	Score for Element	Overall Score
Safety and Health Training		Training							
Emergency Response		First Aid							
Emergency Response		Emergency Preparedness							
Hazard Prevention and Control		Medical Program							
Hazard Prevention and Control		Maintenance							
Hazard Prevention and Control		Hazard Control							
Accident and Record Analysis		Data Analysis							
Accident and Record Analysis		Accident Investigation							
Workplace Analysis		Reporting							
Workplace Analysis		Inspection							
Workplace Analysis		Survey and Hazard Analysis							
Management Leadership and Employee Participation		Contractor Safety							
Management Leadership and Employee Participation		Implementation							
Management Leadership and Employee Participation		Employee Participation							
Management Leadership and Employee Participation		Management Leadership							

Outstanding	5
Superior	4
Basic	3
Developmental	2
Absent or Ineffective	1

OSHA-195 (3/96)

Appendix G

Management Leadership and Employee Participation
Management Leadership

Visible management leadership provides the motivating force for an effective safety and health program. [1989 Voluntary Safety and Health Program Management Guidelines, (b)(1) and (c)(1)]

1	Management demonstrates no policy, goals, objectives, or interest in safety and health issues at this worksite.
2	Management sets and communicates safety and health policy and goals, but remains detached from all other safety and health efforts.
3	Management follows all safety and health rules, and gives visible support to the safety and health efforts of others.
4	Management participates in significant aspects of the site's safety and health program, such as site inspections, incident reviews, and program reviews. Incentive programs that discourage reporting of accidents, symptoms, injuries, or hazards are absent. Other incentive programs may be present.
5	Site safety and health issues are regularly included on agendas of management operations meetings. Management clearly demonstrates--by involvement, support, and example--the primary importance of safety and health for everyone on the worksite. Performance is consistent and sustained or has improved over time.

Management Leadership and Employee Participation
Employee Participation

Employee participation provides the means through which workers identify hazards, recommend and monitor abatement, and otherwise participate in their own protection. [Guidelines, (b)(1) and (c)(1).]

1	Worker participation in workplace safety and health concerns is not encouraged. Incentive programs are present which have the effect of discouraging reporting of incidents, injuries, potential hazards or symptoms. Employees/employee representatives are not involved in the safety and health program.
2	Workers and their representatives can participate freely in safety and health activities at the worksite without fear of reprisal. Procedures are in place for communication between employer and workers on safety and health matters. Worker rights under the Occupational Safety and Health Act to refuse or stop work that they reasonably believe involves imminent danger are understood by workers and honored by management. Workers are paid while performing safety activities.
3	Workers and their representatives are involved in the safety and health program, involved in inspection of work area, and are permitted to observe monitoring and receive results. Workers' and representatives' right of access to information is understood by workers and recognized by management. A documented procedure is in place for raising complaints of hazards or discrimination and receiving timely employer responses.

Management Leadership and Employee Participation
Employee Participation (cont.)

4	Workers and their representatives participate in workplace analysis, inspections and investigations, and development of control strategies throughout facility, and have necessary training and education to participate in such activities. Workers and their representatives have access to all pertinent health and safety information, including safety reports and audits. Workers are informed of their right to refuse job assignments that pose serious hazards to themselves pending management response.
5	Workers and their representatives participate fully in development of the safety and health program and conduct of training and education. Workers participate in audits, program reviews conducted by management or third parties, and collection of samples for monitoring purposes, and have necessary training and education to participate in such activities. Employer encourages and authorizes employees to stop activities that present potentially serious safety and health hazards.

Management Leadership and Employee Participation
Implementation

Implementation means tools, provided by management, that include:
- budget
- information
- personnel
- assigned responsibility
- adequate expertise and authority
- means to hold responsible persons accountable (line accountability)
- program review procedures.

[Guidelines, (b)(1) and (c)(1)]

1	Tools to implement a safety and health program are inadequate or missing.
2	Some tools to implement a safety and health program are adequate and effectively used; others are ineffective or inadequate. Management assigns responsibility for implementing a site safety and health program to identified person(s). Management's designated representative has authority to direct abatement of hazards that can be corrected without major capital expenditure.
3	Tools to implement a safety and health program are adequate, but are not all effectively used. Management representative has some expertise in hazard recognition and applicable OSHA requirements. Management keeps or has access to applicable OSHA standards at the facility, and seeks appropriate guidance information for interpretation of OSHA standards. Management representative has authority to order/purchase safety and health equipment.
4	All tools to implement a safety and health program are more than adequate and effectively used. Written safety procedures, policies, and interpretations are updated based on reviews of the safety and health program. Safety and health expenditures, including training costs and personnel, are identified in the facility budget. Hazard abatement is an element in management performance evaluation.

Appendix G

	Management Leadership and Employee Participation Implementation (cont.)
5	All tools necessary to implement a good safety and health program are more than adequate and effectively used. Management safety and health representative has expertise appropriate to facility size and process, and has access to professional advice when needed. Safety and health budgets and funding procedures are reviewed periodically for adequacy.

	Management Leadership and Employee Participation Contractor Safety
	**Contractor safety**: An effective safety and health program protects all personnel on the worksite, including the employees of contractors and subcontractors. It is the responsibility of management to address contractor safety. [Guidelines, (b)(1) and (c)(1)]
1	Management makes no provision to include contractors within the scope of the worksite's safety and health program.
2	Management policy requires contractor to conform to OSHA regulations and other legal requirements.
3	Management designates a representative to monitor contractor safety and health practices, and that individual has authority to stop contractor practices that expose host or contractor employees to hazards. Management informs contractor and employees of hazards present at the facility.
4	Management investigates a contractor's safety and health record as one of the bidding criteria.
5	The site's safety and health program ensures protection of everyone employed at the worksite, i.e., regular full-time employees, contractors, temporary and part-time employees.

	Workplace Analysis Survey and Hazard Analysis
	**Survey and hazard analysis**: An effective, proactive safety and health program will seek to identify and analyze all hazards. In large or complex workplaces, components of such analysis are the comprehensive survey and analyses of job hazards and changes in conditions. [Guidelines, (c)(2)(i)]
1	No system or requirement exists for hazard review of planned/changed/new operations. There is no evidence of a comprehensive survey for safety or health hazards or for routine job hazard analysis.

	Workplace Analysis Survey and Hazard Analysis (cont.)
2	Surveys for violations of standards are conducted by knowledgeable person(s), but only in response to accidents or complaints. The employer has identified principal OSHA standards which apply to the worksite.
3	Process, task, and environmental surveys are conducted by knowledgeable person(s) and updated as needed and as required by applicable standards. Current hazard analyses are written (where appropriate) for all high-hazard jobs and processes; analyses are communicated to and understood by affected employees. Hazard analyses are conducted for jobs/ tasks/workstations where injury or illnesses have been recorded.
4	Methodical surveys are conducted periodically and drive appropriate corrective action. Initial surveys are conducted by a qualified professional. Current hazard analyses are documented for all work areas and are communicated and available to all the workforce; knowledgeable persons review all planned/changed/new facilities, processes, materials, or equipment.
5	Regular surveys including documented comprehensive workplace hazard evaluations are conducted by certified safety and health professional or professional engineer, etc. Corrective action is documented and hazard inventories are updated. Hazard analysis is integrated into the design, development, implementation, and changing of all processes and work practices.

	Workplace Analysis Inspection
Inspection: To identify new or previously missed hazards and failures in hazard controls, an effective safety and health program will include regular site inspections. [Guidelines, (c)(2)(ii)]	
1	No routine physical inspection of the workplace and equipment is conducted.
2	Supervisors dedicate time to observing work practices and other safety and health conditions in work areas where they have responsibility.
3	Competent personnel conduct inspections with appropriate involvement of employees. Items in need of correction are documented. Inspections include compliance with relevant OSHA standards. Time periods for correction are set.
4	Inspections are conducted by specifically trained employees, and all items are corrected promptly and appropriately. Workplace inspections are planned, with key observations or check points defined and results documented. Persons conducting inspections have specific training in hazard identification applicable to the facility. Corrections are documented through follow-up inspections. Results are available to workers.
5	Inspections are planned and overseen by certified safety or health professionals. Statistically valid random audits of compliance with all elements of the safety and health program are conducted. Observations are analyzed to evaluate progress.

Appendix G

Workplace Analysis Hazard Reporting	
A reliable hazard reporting system enables employees, without fear of reprisal, to notify management of conditions that appear hazardous and to receive timely and appropriate responses. [Guidelines, (c)(2)(iii)]	
1	No formal hazard reporting system exists, or employees are reluctant to report hazards.
2	Employees are instructed to report hazards to management. Supervisors are instructed and are aware of a procedure for evaluating and responding to such reports. Employees use the system with no risk of reprisals.
3	A formal system for hazard reporting exists. Employee reports of hazards are documented, corrective action is scheduled, and records maintained.
4	Employees are periodically instructed in hazard identification and reporting procedures. Management conducts surveys of employee observations of hazards to ensure that the system is working. Results are documented.
5	Management responds to reports of hazards in writing within specified time frames. The workforce readily identifies and self-corrects hazards; they are supported by management when they do so.

Accident and Record Analysis Accident Investigation	
Accident investigation: An effective program will provide for investigation of accidents and "near miss" incidents, so that their causes, and the means for their prevention, are identified. [Guidelines, (c)(2)(iv)]	
1	No investigation of accidents, injuries, near misses, or other incidents is conducted.
2	Some investigation of incidents takes place, but root cause may not be identified, and correction may be inconsistent. Supervisors prepare injury reports for lost time cases.
3	OSHA-101 is completed for all recordable incidents. Reports are generally prepared with cause identification and corrective measures prescribed. (Author's note: Now OSHA-301.)
4	OSHA-recordable incidents are always investigated, and effective prevention is implemented. Reports and recommendations are available to employees. Quality and completeness of investigations are systematically reviewed by trained safety personnel.
5	All loss-producing accidents and "near-misses" are investigated for root causes by teams or individuals that include trained safety personnel and employees.

Accident and Record Analysis
Data Analysis

Data analysis: An effective program will analyze injury and illness records for indications of sources and locations of hazards, and jobs that experience higher numbers of injuries. By analyzing injury and illness trends over time, patterns with common causes can be identified and prevented. [Guidelines, (c)(2)(v)]

1	Little or no analysis of injury/illness records; records (OSHA 200/101, exposure monitoring) are kept or conducted. (Author's note: Now OSHA 300/301.)
2	Data is collected and analyzed, but not widely used for prevention. OSHA - 101 (Author's note: Now OSHA-301) is completed for all recordable cases. Exposure records and analyses are organized and are available to safety personnel.
3	Injury/illness logs and exposure records are kept correctly, are audited by facility personnel, and are essentially accurate and complete. Rates are calculated so as to identify high risk areas and jobs. Workers compensation claim records are analyzed and the results used in the program. Significant analytical findings are used for prevention.
4	Employer can identify the frequent and most severe problem areas, the high risk areas and job classifications, and any exposures responsible for OSHA recordable cases. Data are fully analyzed and effectively communicated to employees. Illness/injury data are audited and certified by a responsible person.
5	All levels of management and the workforce are aware of results of data analyses and resulting preventive activity. External audits of accuracy of injury and illness data, including review of all available data sources are conducted. Scientific analysis of health information, including non-occupational data bases is included where appropriate in the program.

Hazard Prevention and Control
Hazard Control

Hazard Control: Workforce exposure to all current and potential hazards should be prevented or controlled by using engineering controls wherever feasible and appropriate, work practices and administrative controls, and personal protective equipment (PPE). [Guidelines, (c)(3)(i)]

1	Hazard control is seriously lacking or absent from the facility.
2	Hazard controls are generally in place, but effectiveness and completeness vary. Serious hazards may still exist. Employer has achieved general compliance with applicable OSHA standards regarding hazards with a significant probability of causing serious physical harm. Hazards that have caused past injuries in the facility have been corrected.

	Hazard Prevention and Control Hazard Control (cont.)
3	Appropriate controls (engineering, work practice, and administrative controls, and PPE) are in place for significant hazards. Some serious hazards may exist. Employer is generally in compliance with voluntary standards, industry practices, and manufacturers' and suppliers' safety recommendations. Documented reviews of needs for machine guarding, energy lockout, ergonomics, materials handling, bloodborne pathogens, confined space, hazard communication, and other generally applicable standards have been conducted. The overall program tolerates occasional deviations.
4	Hazard controls are fully in place, and are known and supported by the workforce. Few serious hazards exist. The employer requires strict and complete compliance with all OSHA, consensus, and industry standards and recommendations. All deviations are identified and causes determined.
5	Hazard controls are fully in place and continually improved upon based on workplace experience and general knowledge. Documented reviews of needs are conducted by certified health and safety professionals or professional engineers, etc.

	Hazard Prevention and Control Maintenance
	**Maintenance:** An effective safety and health program will provide for facility and equipment maintenance, so that hazardous breakdowns are prevented. [Guidelines, (c)(3)(ii)]
1	No preventive maintenance program is in place; break-down maintenance is the rule.
2	There is a preventive maintenance schedule, but it does not cover everything and may be allowed to slide or performance is not documented. Safety devices on machinery and equipment are generally checked before each production shift.
3	A preventive maintenance schedule is implemented for areas where it is most needed; it is followed under normal circumstances. Manufacturers' and industry recommendations and consensus standards for maintenance frequency are complied with. Breakdown repairs for safety related items are expedited. Safety device checks are documented. Ventilation system function is observed periodically.
4	The employer has effectively implemented a preventive maintenance schedule that applies to all equipment. Facility experience is used to improve safety-related preventative maintenance scheduling.
5	There is a comprehensive safety and preventive maintenance program that maximizes equipment reliability.

Hazard Prevention and Control Medical Program	
	An effective safety and health program will include a suitable medical program where it is appropriate for the size and nature of the workplace and its hazards. [Guidelines, (c)(3)(iv)]
1	Employer is unaware of, or unresponsive to medical needs. Required medical surveillance, monitoring, and reporting are absent or inadequate.
2	Required medical surveillance, monitoring, removal, and reporting responsibilities for applicable standards are assigned and carried out, but results may be incomplete or inadequate.
3	Medical surveillance, removal, monitoring, and reporting comply with applicable standards. Employees report early signs/symptoms of job-related injury or illness and receive appropriate treatment.
4	Health care providers provide follow-up on employee treatment protocols and are involved in hazard identification and control in the workplace. Medical surveillance addresses conditions not covered by specific standards. Employee concerns about medical treatment are documented and responded to.
5	Health care providers are on-site for all production shifts and are involved in hazard identification and training. Health care providers periodically observe the work areas and activities and are fully involved in hazard identification and training.

Emergency Response Emergency Preparedness	
	**Emergency preparedness**: There should be appropriate planning, training/drills, and equipment for response to emergencies. Note: In some facilities the employer plan is to evacuate and call the fire department. In such cases, only applicable items listed below should be considered. [Guidelines, (c)(3)(iii) and (iv)]
1	Little or no effective effort to prepare for emergencies.
2	Emergency response plans for fire, chemical, and weather emergencies as required by 29 CFR 1910.38, 1910.120, or 1926.35 are present. Training is conducted as required by the applicable standard. Some deficiencies may exist.
3	Emergency response plans have been prepared by persons with specific training. Appropriate alarm systems are present. Employees are trained in emergency procedures. The emergency response extends to spills and incidents in routine production. Adequate supply of spill control and PPE appropriate to hazards on site is available.
4	Evacuation drills are conducted no less than annually. The plan is reviewed by a qualified safety and health professional.

Emergency Response
Emergency Preparedness (cont.)

5	Designated emergency response team with adequate training is on-site. All potential emergencies have been identified. Plan is reviewed by the local fire department. Plan and performance are reevaluated at least annually and after each significant incident. Procedures for terminating an emergency response condition are clearly defined.

Emergency Response
First Aid

First aid/emergency care should be readily available to minimize harm if an injury or illness occurs. [Guidelines, (c)(3)(iii) and (iv)]

1	Neither on-site nor nearby community aid (e.g., emergency room) can be ensured.
2	Either on-site or nearby community aid is available on every shift.
3	Personnel with appropriate first aid skills commensurate with likely hazards in the workplace and as required by OSHA standards (e.g., 1910.151, 1926.23) are available. Management documents and evaluates response time on a continuing basis.
4	Personnel with certified first aid skills are always available on-site; their level of training is appropriate to the hazards of the work being done. Adequacy of first aid is formally reviewed after significant incidents.
5	Personnel trained in advanced first aid and/or emergency medical care are always available on-site. In larger facilities a health care provider is on-site for each production shift.

Safety and Health Training

Safety and health training should cover the safety and health responsibilities of all personnel who work at the site or affect its operations. It is most effective when incorporated into other training about performance requirements and job practices. It should include all subjects and areas necessary to address the hazards at the site. [Guidelines, (b)(4) and (c)(4)]

1	Facility depends on experience and peer training to meet needs. Managers/supervisors demonstrate little or no involvement in safety and health training responsibilities.
2	Some orientation training is given to new hires. Some safety training materials (e.g., pamphlets, posters, videotapes) are available or are used periodically at safety meetings, but there is little or no documentation of training or assessment of worker knowledge in this area. Managers generally demonstrate awareness of safety and health responsibilities, but have limited training themselves or involvement in the site's training program.

	Safety and Health Training (cont.)
3	Training includes OSHA rights and access to information. Training required by applicable standards is provided to all site employees. Supervisors and managers attend training in all subjects provided to employees under their direction. Employees can generally demonstrate the skills/knowledge necessary to perform their jobs safely. Records of training are kept and training is evaluated to ensure that it is effective.
4	Knowledgeable persons conduct safety and health training that is scheduled, assessed, and documented, and addresses all necessary technical topics. Employees are trained to recognize hazards, violations of OSHA standards, and facility practices. Employees are trained to report violations to management. All site employees--including supervisors and managers--can generally demonstrate preparedness for participation in the overall safety and health program. There are easily retrievable scheduling and record keeping systems.
5	Knowledgeable persons conduct safety and health training that is scheduled, assessed, and documented. Training covers all necessary topics and situations, and includes all persons working at the site (hourly employees, supervisors, managers, contractors, part-time and temporary employees). Employees participate in creating site-specific training methods and materials. Employees are trained to recognize inadequate responses to reported program violations. Retrievable record keeping system provides for appropriate retraining, makeup training, and modifications to training as the result of evaluations.

Appendix G

Bibliography

Author's note: The references marked with asterisks concern the OSHA Voluntary Protection Programs, safety and health management practices to attain excellence, strategic planning, and leadership. The other references are excellent sources for general management, leadership principles, and practices.

Adams, Edward E. *Total Quality Safety Management—An Introduction.* Des Plaines, IL: American Society of Safety Engineers. 1995.

Blanchard, Ken, John P. Carlos, and Alan Randolph. *Empowerment takes more than a minute.* San Francisco: Berrett-Keohler, 1966.

Blanchard, Kenneth and Norman Vincent Peale. *The Power of Ethical Management.* New York: Ballantine Books, 1988.

Bryson, John M. *Strategic Planning for Public and Nonprofit Organizations: A Guide to Strengthening and Sustaining Organizational Achievement.* San Francisco: Jossey-Bass Inc., Publishers, 1988.

Coates, Joseph F., Vary T. Coates, and Jennifer Heinz Jarratt. *Issues Management: How You Can Plan, Organize and Manage for the Future,* Mt. Airy, MD: Lomond Publications, Inc., 1986.

*Covey, Stephen R. *The 7 habits of highly effective people: restoring the character ethic.* New York: Simon & Schuster, 1989.

*Covey, Stephen R. *Principle-centered leadership.* New York: Summit Books, 1991.

*Covey, Stephen R., A. Roger Merrill, and Rebecca R. Merrill. *First things first: to live, to learn, to leave a legacy.* New York: Simon & Schuster, 1994.

Empowerment Workbook, 1993 Workshop Notes. Dallas: Covey Leadership Center. 1992.

*Deming, W. Edwards. *Out of the crisis.* Cambridge, MA: MIT Press, 1986.

*De Pree, Max. *Leadership is an Art.* New York: Doubleday, 1989.

Drucker, Peter F. *The Effective Executive.* New York: Harper & Row, 1967.

Drucker, Peter F. *The Practice of Management.* New York: Harper & Row, 1954.

Drucker, Peter F. *The Age of Discontinuity: Guidelines to our changing society.* New York: Harper & Row, 1969.

Drucker, Peter F. *Management Challenges for the 21st Century.* New York: HarperCollins, 1999.

Environmental Protection Magazine. Dallas: Stevens Publishing Company.

*Garner, Charlotte A. and Patricia O. Horn. *How Smart Managers Improve Their Safety and Health Systems: Benchmarking with OSHA VPP Criteria.* Des Plaines, IL: American Society of Safety Engineers, 1999.

*Goodstein, Leonard, Timothy Nolan, and J. William Pfeiffer. *Applied Strategic Planning: A comprehensive guide.* New York: McGraw-Hill, 1993.

Haltenhoff, C. Edwin. *The CM Contracting System: Fundamentals and Practices.* Upper Saddle River, NJ: Prentice Hall, 1999.

Hammer, Michael and Steven A. Stanton. *The Reengineering Revolution.* New York: HarperCollins, 1995.

*Hartnett, John. *OSHA in the Real World: How to maintain workplace safety while keeping your competitive edge.* Santa Monica, CA: Merritt Publishing, 1996.

Hesselbein, Frances, Marshall Goldsmith; and Richard Beckhard, editors. *The Organization of the Future.* San Francisco: Jossey-Bass Inc., Publishers, 1997.

Hinze, Jimmie W. *Construction Safety.* Upper Saddle River, NJ: Prentice Hall, Inc., 1997.

Hislop, Richard D. *Construction Site Safety: A Guide for Managing Contractors.* Boca Raton, FL: CRC Press LLC, 1999.

The Houston Chronicle, December 1996–January 1997.

Hunt, V. Daniel. *Quality in America: How to implement a competitive quality program.* Chicago: Irwin Professional Publishing, 1992.

Injury Facts. Itasca, IL: National Safety Council.

Kase, Donald W. and Kay J. Wiese. *Safety Auditing: A Management Tool.* New York: Van Nostrand Reinhold, 1990.

*Kotter, John P. *Leading Change.* Boston: Harvard Business School Press, 1996.

Krause, Thomas R., John H. Hidley, and Stanley J. Hodson. *The Behavior-Based Safety Process: Managing involvement for an injury-free culture.* New York: Van Nostrand Reinhold, 1990.

The Leader Magazine. Falls Church, VA: The Voluntary Protection Programs Participants' Association.

*Lebow, Rob and William L. Simon. *Lasting Change: The Shared Values Process That Makes Companies Great*. New York: Van Nostrand Reinhold, 1997.

Malcolm Baldrige National Quality Award. Gaithersburg, MD: National Institute of Standards and Technology.

McSween, Terry E. *The Values-Based Safety Process: Improving Your Safety Culture with a Behavioral Approach*. New York: Van Nostrand Reinhold, 1995.

The New York Times, December 29, 1996.

Oakley, Ed and Doug Krug. *Enlightened Leadership: Getting to the Heart of Change*. New York: Simon & Schuster, 1991.

Occupational Health & Safety Magazine. Dallas: Stevens Publishing Company.

*Occupational Safety and Health Administration, US Department of Labor, Washington, DC.

Professional Safety, Journal of the American Society of Safety Engineers. Des Plaines, IL.

Rice, Craig S. *Strategic Planning for the Small Business: Situations, Weapons, Objectives & Tactics*. Holbrook, MA: Adams Media Corporation, 1990.

*Richardson, Margaret R. *Managing Worker Safety and Health for Excellence*. New York: Van Nostrand Reinhold, 1997.

*Richardson, Margaret R. *Preparing for the Voluntary Protection Programs: Building Your Star Program*. New York: John Wiley & Sons, Inc., 1999.

Swartz, George, editor. *Safety Culture and Effective Safety Management*. Itasca, IL: National Safety Council, 2000.

*Terrell, Milton J. *Safety and Health Management in the Nineties: Creating a Winning Program*. New York: Van Nostrand Reinhold, 1995.

Tracy, Diane. *10 Steps to Empowerment: A Common-Sense Guide to Managing People*. New York: William Morrow and Company, Inc., 1990.

Walton, Mary. *The Deming Management Method*. New York: Putnam, 1986.

Woodward, Harry and Steve Buchholz. *Aftershock: Helping People Through Corporate Change.* New York: John Wiley & Sons, 1987.

*Zimmerman, John, with Benjamin Tregoe. *The Culture of Success: Building a Sustained Competitive Advantage by Living Your Corporate Beliefs.* New York: McGraw-Hill, 1997.

Index

-A-

Abundance mentality, 132, 137
Accident investigation, 47, 114, 119, 138
American Industrial Hygiene Association (AIHA), 24
Americans with Disabilities Act, 6–7
Annual program evaluation, 40, 45, 58, 64, 68, 75, 85, 86, 102, 109, 111–27, 165, 168, 175, 180, 188, 192. *See also:* Safety, Health, and Environmental program(s)
    post-audit actions, 124–27
    reviews, 112-13
    self-evaluation document, 124
    worksite analysis, 117–19. *See also:* Job hazard analysis
Applied strategic planning, 62, 66, 93, 95, 97, 101

-B-

Behavior-based training, 120
Benchmarking, 83–85, 127, 166
Blank, Arthur, 181
Brennecke, Robert A., 24–25
Brown, Stephen, 130, 137, 139, 140–41
Budget
    for safety and health expenditures, 40, 102, 126, 168, 176
    and the strategic plan, 65, 73, 76, 92–93
Bureau of Labor Statistics (BLS), 5
    Census of Fatal Occupational Injuries, 5
Business Know-How, 27

-C-

CareerBuilder, 27
Certified Industrial Hygienist (CIH), 116
Certified Safety Professional (CSP), 116
Chandler, Phillip B., 28
Change/innovation within companies, 33–34, 154
    dynamics of, 165–66, 182–83
    overcoming obstacles to, 82–83
    resistance to, 72
Chronicle (Houston, Texas), 4
Communicating with employees, 41–43, 66, 73, 97, 101, 115, 116, 148, 152, 169–70, 181–182
    creating a sense of urgency, 154–55, 160, 179–80
    methods, 155–63
    plan for, 153–55
Compliance to OSHA Standards, 3, 8, 15, 46, 67, 126, 154, 165, 189. *See also:* Occupational Safety and Health Act (OSH Act) *and* Occupational Safety and Health Administration (OSHA)
    setting priorities and, 92–93
Computer-based training, 190
Conoco, Incorporated, 166–67, 182
Consent decree, 6
Construction Safety and Health Program, General Safety and Health Provisions (CFR 1926.20), 8
Contingency plan, 8, 28, 73, 85–86, 93–95
Corporate culture, 57, 63, 78, 99, 154, 179, 192
    changing the dynamic of, 146–52, 163, 188
    examining the, 79–82,
    and gap analysis, 64
    strategic planning and, 72–75
Covey, Stephen, 132, 136, 143, 151, 162, 169, 185, 193

Critical Element Schedule, 84, 98

-D-

Dear, Joseph, 111
Decision-making process
    characteristics of an effective, 55–58
    flow diagram, 56
Deming, W. Edwards, 145, 146
Department of Justice (DOJ), 6, 16
Department of Transportation (DOT), 9, 16, 96
De Pree, Max, 136, 152
Drucker, Peter, F., 33, 53, 54, 55, 57, 58, 89, 129, 145

-E-

Economy of the future, 188–89
Emergency response program, 48, 95, 122, 123, 167
Employee involvement, 34, 58, 82, 89, 98, 113, 115, 129–32, 135–37, 139–42, 148
    management leadership and, 38–46
    meaningful, 131
Environmental compliance, 9–10, 54–55
    costs, 21–22
Environmental Protection Agency (EPA), 8–10, 16, 96
    Risk Management Program (RMP), 8
Ergonomics, 10–11, 121, 189
    survey, 66
Experience modifier rate (EMR), 104
ExxonMobil, 79

-F-

Farabaugh, Peggy, 12
Fulwiler, Richard D., 25

-G-

Gap analysis, 30, 48, 50–68, 73, 75, 78, 94, 102, 112, 162, 166, 169, 175
    10-step procedure, 61–68
    action plan, 58–61
    compliance worksheet, 58–59, 66
    and effective decision-making, 55–58
    worksheet, 59
Geon Company, 174
Goodstein, Leonard, 53, 71, 89
Guimond, Rich, 37

-H-

Hammer, Michael, 98, 126, 161
Handy, John, 190
Hinze, Jimmie, 25–26
Holzapfel, Richard, 126
Home Depot, 181
Horn, Patricia, 133, 187

-I-

Incentives and recognition programs, 103–104, 107, 176
Industrial hygiene surveys, 117, 122
In Pursuit of Excellence, 145
Injuries
    indirect costs of, 25
    trend of occupational illness and, 24, 27
    zero workplace, 41
International Brotherhood of Electrical Workers (IBEW), 139, 142
International Organization for Standardization (ISO), 11
    environmental management system model (ISO 14001), 11
    quality assurance system (ISO 9001), 11

–J–

Job hazard analysis, 38, 46–48, 85, 89, 98, 101, 118
Job insecurity, 26–27
Job safety analysis (JSA), 41–42, 118
Job-specific training, 2, 165
Johnson Space Center (JSC), 65–66, 69, 86, 90, 126, 133, 145, 148, 151, 159,
    communication methods and practices, 155–57, 161–62
    initial implementation of the VPP plan at, 101–106
    gap analysis action plan, 58–61
    safety and health program survey, 79-82
    safety and total health day, 176–78
    strategic plan, 73–82

–K–

Kotter, John, 178

–L–

Leadership
    commitment, 129–36
    role of corporate, 102–104, 108, 109, 148–52, 154, 160–61, 181, 192
Leadership Is An Art, 136
Leading Change, 178
Lesko, Bill, 174

–M–

Maderos, Manuel A., 139, 141–42
Management By Objectives (MBO), 135
Marcus, Bernard, 181
Mission statement, 48, 76–79, 125, 145, 148, 150, 152–54, 188
Mobil Oil Company, 28, 137, 171
Monsanto Chemical Company, 28
Motorola, Incorporated, 28–29, 37

–N–

NASA, 81, 82, 86, 103, 145
National News Report, 124, 125
National Safety Council (NSC), 22–24
Injury Facts, 22–24
Nolan, Timothy, 53, 71, 89
North American Industry Classification System (NAICS), 147

–O–

Occupational health care program, 47
Occupational Health and Safety magazine, 12
Occupational Safety and Health Act (OSH Act), 7, 10, 16, 44, 146, 189
    general duty clause, 10
Occupational Safety and Health Administration (OSHA), 7, 16, 28, 37, 45, 50, 94–97, 130, 186. *See also:* OSHA Standard(s); Voluntary Protection Program(s) (VPP); *and* VPP management system criteria
    compliance officer(s), 3, 9, 102, 124, 154
    compliance to standards, 3, 8, 15, 46, 67, 92–93, 126, 154, 165, 189
    ergonomics strategy, 10–11
    fines/penalties, 5, 16, 20, 31
    Hispanic outreach, 11
    injury/illness rate targeting system and cooperative compliance program (CCP), 146–47
    inspections, 4, 16, 20–21, 96, 97, 147
    Multi-Employers Site Citation Policy Directive (CPL 2-0.124), 7
    National Emphasis Programs (NEPs), 10
    response to serious accidents and fatalities at VPP sites, 21
    self-assessment checklist for

model SH&E management system, 39–46
site-specific targeting (SST), 147
Strategic Management Plan for FY2004–2008, 20–21
training, 2, 11, 190
Organization of the future, 185–93
OSHA Cooperative Consultation Office, 67, 94, 131
OSHA Facts, 19, 24
OSHA Log, 49, 96, 138–40, 147
OSHA Office of Compliance, 94
OSHA permissible exposure limits (PELs), 117
OSHA Standard(s), 3, 8–9
    incorporated into computer software, 25, 191
    Noise Standard, 7
    Process Safety Management of Highly Hazardous Chemicals Rule (29 CFR 1910.119), 8
    Standards Interpretations Memorandum, 8
OSHA-VPP Requirements for Construction Site Joint Labor-Management Committee for Safety and Health, 138–39

–P–

Performance-based program, 8
Personal protective equipment, 2, 39, 47, 115, 120, 123–24
Peters, Tom, 97, 145
Pfeiffer, William J., 53, 71, 89
Potlatch Corporation, 130, 137, 139
Principle-Centered Leadership, 169, 185
Proctor & Gamble, 25

–R–

Reengineering, definition, 126
Revisions to the Voluntary Protection Programs to Provide Safe and Healthful Working Conditions, 40, 118

Richardson, Peggy, 187
Root cause analysis, 124

–S–

Safety, Health, and Environmental (SH&E) program(s). *See also:* Annual program evaluation *and* SH&E management systems
    assessing a current program. *See:* SH&E program assessment
    budgeting for, 75, 102, 109, 115
    communicating with employees, 41–43, 66, 73, 97, 101, 115, 116, 148, 151–63, 169–70, 179–82
    contingency plans. *See:* Strategic plan
    continuing improvement process, 166–83
    contract workers, 44, 116–17
    employee orientation process, 174
    environmental compliance, 9–10, 54–55
    health care services, 121–22
    incentives/recognition in, 101, 104, 107, 108–109, 149, 176–79, 182
    legal challenges related to, 6–10, 15–16
    maintaining employee interest, 171–78
    mission statement, 48, 76–79, 125, 145, 148, 150, 152–54, 188
    organizing a steering committee, 37, 76–77, 154, 174–75, 182
    Phase One Project Management Plan, 74, 91
    process safety management (PSM), 121
    Program Development Worksheet, 59–60, 66, 68

Index

questions regarding, 2–3, 19, 35, 53–54, 62–63, 78
reactive type, 3–5
teaching the leadership style, 67
training, 48, 122–23, 150
unifying focus, 75–76
value of, 1–5, 13–14, 19–20, 35
values-based system, 148, 150
written policies and procedures manual, 41, 116
$AFETY PAYS software program, 25
Scarcity mentality, 132
Shell Chemicals, 112
SH&E management systems
    characteristics of, 12–13
    implementing the model, 68, 89–109, 170, 187. *See also:* Voluntary Protection Programs (VPP)
    integrating action plans, 89–91, 98, 109, 170
    internal control, 28–29, 31
    role of corporate leadership, 102–104, 108, 109, 148–52, 154, 160–61, 181, 192
    role of the steering committee, 99, 109, 161
    versus traditional safety and health programs, 34–35
    worksite analysis elements, 117–22, 124
SH&E program assessment, 2–3, 30, 33, 50, 75, 83–84, 94, 98, 127, 166. *See also:* Gap analysis
    four critical elements, 38–48
Smith, Gregory P., 27
Standard Industrial Classification (SIC), 147
Stanton, Steven A., 98, 126, 161
Star Program requirements, 114
Strategic plan, 28, 63, 64, 69, 83, 85–86, 92–93, 99, 100, 142, 169, 185
    budgeting and the, 65, 73, 76, 92–93
    and corporate culture, 72–75

gap analysis and the, 73
mission statement and the, 76–79
Survey, JSC safety and health, 79–82

–T–

The Age of Discontinuity, 53
The Culture of Success, 176, 181
The Effective Executive, 54, 58
The New York Times, 26
The Leader, 15, 172–173
Toolbox meetings, 116
Tregoe, Benjamin B., 176, 181

–U–

Union agreement, 133

–V–

Values survey, 63, 97–98, 148
Vision within a company, 30, 71
Voluntary Protection Program Participants Association (VPPPA), 30, 77, 101, 129, 137, 139, 141, 161, 171–72
Voluntary Protection Programs (VPP), 8, 11, 19, 20, 28, 31, 35–36, 42, 44, 45, 48–50, 73, 95, 185, 192
    accountability element, 65, 73, 98, 102, 135–36
    codification of, 8, 14–15, 86
    communication lines, 191
    continual improvement, 30, 68, 73, 186
    cost to implement, 29–30, 31, 187
    criteria as benchmarks, 83–85, 127, 166
    management system criteria. *See:* VPP management system criteria
    mission statement, 125, 148
    performance appraisal system, 115–16

Policies and Procedures Manual, 2, 113–14, 131, 133
Program Evaluation Profile (PEP), 127, 192
self-evaluation, 111–13, 116, 124–25
Star program requirements, 114
union participation, 45–46, 133
using consultants, 126
VPP management system criteria, 50, 53–54, 58, 61, 62, 76, 82, 91, 93, 97, 100, 125, 136, 146, 167, 182, 190
    benchmarking with, 127–28
    employee involvement, 34, 38–39, 41–45, 58, 82, 89, 98, 115, 129–32, 135, 136–37, 139–42, 148
    hazard analysis and control, 38, 46–48, 85, 89, 98, 101
    management leadership, 38–46, 78, 89, 113, 115, 130–36
    safety and health training, 38, 48, 89, 98
    self-assessment, 83–85

–W–

Worker confidence/trust, 26–27, 149, 160, 175
Workers' compensation, 16, 22–26, 28, 96, 186–87

–Z–

Zimmerman, Sr., John, 176, 181